CRAIG ROBERTSON

THE FILING CABINET
A VERTICAL HISTORY OF INFORMATION

University of Minnesota Press
Minneapolis | London

Portions of this book were published in "File" in *Uncertain Archives: Critical Keywords for Big Data,* ed. Nanna Bonde Thylstrup, Daniela Agostinho, Annie Ring, Catherine D'Ignazio, and Kristin Veel (Cambridge, Mass.: MIT Press, 2021). Portions of chapters 5 and 6 were published in a different form in "Learning to File: Reconfiguring Information and Information Work in the Early Twentieth Century," *Technology and Culture* 58, no. 4 (2017): 955–81; copyright 2017 Society for the History of Technology; reprinted with permission from Johns Hopkins University Press.

Copyright 2021 by the Regents of the University of Minnesota

All rights reserved. No part of this publication may be reproduced, stored in a retrieval system, or transmitted, in any form or by any means, electronic, mechanical, photocopying, recording, or otherwise, without the prior written permission of the publisher.

Published by the University of Minnesota Press
111 Third Avenue South, Suite 290
Minneapolis, MN 55401-2520
http://www.upress.umn.edu

ISBN 978-1-5179-0945-1 (hc)
ISBN 978-1-5179-0946-8 (pb)
A Cataloging-in-Publication record for this book is available from the Library of Congress.

Printed in the United States of America on acid-free paper

The University of Minnesota is an equal-opportunity educator and employer.

For E & e

Contents

Preface ix

Introduction: The Efficient Work of Paperwork 1

PART I
THE CABINET

1. Verticality: A Skyscraper for the Office 31
2. Integrity: Paper's Steel Enclosure 63
3. Cabinet Logic: Efficiency through Partitions 103

PART II
FILING

4. Granular Certainty: Applying System to the Office 131
5. Automatic Filing: Memory for the Machine 161
6. The Ideal File Clerk: Controlling Gender in the Office 195
7. Domestic Storage: Cabinet Logic in the Home 223

Afterword: Out of Time, Out of Place 251

Acknowledgments 261
Notes 265
Index 301

Preface

This project came out of the research for my previous book on the history of the U.S. passport—from the act of researching, not from the book's content.[1] Having spent weeks at the National Archives at College Park in Maryland, struggling through thousands of reels of unindexed microfilm records of nineteenth-century U.S. diplomatic correspondence looking for references to passports, I arrived at the records for 1906. That year, the State Department adopted a numerical filing system. Suddenly, every American diplomatic office began using the same number for passport correspondence, with decimal numbers subdividing issues and cases. Rather than scrolling through microfilm images of formerly bound pages organized in chronological order, I could go straight to passport-relevant information that had been gathered in one place. With subject trumping chronology, my research changed dramatically. I was now able to find the papers I wanted with a lot less effort.

I subsequently discovered that I had Elihu Root to thank for making my research easier. A lawyer with clients such as Andrew Carnegie, Root became secretary of state in 1905, having served a previous administration as secretary of war, during which time he is credited with modernizing the U.S. Army. When he arrived at the State Department he described himself as "a man trying to conduct the business of a large metropolitan law-firm in the office of a village squire."[2] The department's record-keeping practices contributed to his frustration. As was common practice in American offices through the nineteenth century, clerks used press books or copybooks to store incoming and outgoing correspondence in chronologically ordered bound volumes with limited indexing. The tipping point for Root came when a request he made for a handful of letters resulted in several large volumes appearing on his desk. In response, he demanded that a vertical filing system be adopted. As I was well aware, in 1906 the department began to use a numerical subject-based filing system housed in vertical filing cabinets; a more comprehensive decimal filing system followed in 1910.[3]

My curiosity led to initial research that revealed that many people at the beginning of the twentieth century shared my excitement about the introduction of a decimal filing system. Frustration with the use of bound volumes for storing paper provided the motivation for many individuals and institutions to adopt filing systems using unbound paper in the decades before and after the State Department did. That transition tended to be expressed as a move from bound books to filing cabinets, and before I knew it I was starting to research a history of the filing cabinet.

The shift in focus from filing systems to filing cabinets means the book you are about to read is a history of storage, not a history of classification. To focus on storage is to bring the filing cabinet to the foreground, to see how the cabinet as an object shapes people's interaction with information rather than explore the effects of systems of classification on knowledge. This is a history about the tabs that make classification and indexing systems visible, not a history that uses the filing cabinet to explore the consequences of organizing papers by author name or subject. Classification and indexing (and the blurring of those in the early twentieth century) do appear in this history, but as ways to rethink storage and to understand the filing cabinet's role in the development of a modern conception of information.

As I thought and wrote my way through the discovery that I was writing a history of storage, I had a lot of opportunities to talk about the project with various people. Those who provided useful feedback and information are noted in my acknowledgments. But there were others who were less helpful. As is the way of the humanities, at different moments a number of people told me that the book I wanted to research—was researching, was writing, was revising for publication—had already been written. This "advice" usually came packaged in the assured confidence of individuals who had neither heard me present my work nor read it. How could I not know the work of business historian JoAnne Yates or the late media studies scholar Cornelia Vismann? Didn't I realize they had already written my book?

To reassure myself that I was indeed writing a different book, I took advantage of the fact that I live in the city where Yates works and made an appointment to see her. A scholar as gracious as she is smart, Yates assured me I was not writing her book (as she noted, the filing cabinet is only one of many objects she discusses in *Control through Communication*).[4] More important, Yates also noted that many of the concepts and academic debates I was bringing to the filing cabinet were not around when she researched and wrote her book thirty years earlier. She encouraged me to continue writing the book. Subsequently, she sent me a box that she discovered while cleaning out her house—it contained handwritten notes and photocopies from her research for *Control through Communication*.

One of the concepts that Yates referred to in our conversation was "materiality." There have been many "material turns" in academic disciplines and fields in recent decades. My home field of media studies is no exception, although as a field defined by the examination of the materiality of communication, it has been trading in versions of these debates since its inception. However, the longer I spent thinking about the filing cabinet, the less important materiality became as an organizing concept. That did not stop me sticking for too long with a three-part structure for the book, each part focused on one of what I consider to be the three fundamental aspects of materiality: physical object (the filing cabinet), socioeconomic context (efficiency), and use (file clerks). Finally, as I began to revise the manuscript, I recognized that direct and consistent theoretical engagement with the stakes and issues raised by materiality was not central to the book I was actually writing—a book located somewhere between critical media studies and history, lacking the direct theoretical engagement the former leans toward and possibly having too much theoretical engagement for the tastes of many who favor the latter.

Despite my ambivalence about exploring the concept of materiality, I remain interested in the filing cabinet primarily as an object, as a technology. In this book the filing cabinet is not a passive object. When we use a filing cabinet, the possibilities available to us are not unlimited. I want to get inside the filing cabinet to understand how it works and the limits it places on what it means to store and to retrieve. This approach borrows from the initial move in recent media materialism, the idea that to study media we need to understand the realities of science and engineering. However, rather than use this perspective to focus on computers (and circuits, hardware, and voltage differences), I look at drawer slides, folders, tabs, and steel to appreciate how the filing cabinet works.

The turn to materiality in media studies (and the technical turn in the humanities in general) has produced good and bad work in equal measure. The latter tends to appear when media materialism becomes thick description that rivals Clifford Geertz's essay on Balinese cockfights. I choose that example deliberately, not only because it is canonical both within and outside anthropology but also because a lot of the work in media studies that gives materiality a bad name is easily caricatured as boys talking about their toys.[5]

My frustration with thick description and media studies manifested itself in a paper I delivered in 2015 at the annual conference of the Society for Cinema and Media Studies. My breakthrough in writing that paper occurred with the realization that if other scholars expected me to care about their detailed descriptions of small corners of the Internet or their favorite video games, then I would treat people to a step-by-step description of how the filing cabinet works. Some of the audience clearly appreciated the paper as parody; elements of that appear in chapter 3 of this book when I draw on that presentation, which I hope readers will similarly appreciate. For me, writing that paper, and the chapter, was enjoyable because it allowed me to revel in the absurdity of describing in detail how a follower block works, how folders store paper, and why designing the perfect tab was so difficult. However, both the presentation and the chapter ultimately fail as parody because I don't want people to be left asking, What does this mean? Therefore, along with detailed description I attempt to explain why I think the description matters, to be as clear as possible about why it is important to understand how the drawer slides in a filing cabinet work.

In explaining the significance of the workings of a filing cabinet, my focus is on culture and social relations of power rather than on direct engagement with theoretical debates. This is where I see an essential difference between this book and Cornelia Vismann's landmark work *Files: Law and Media Technology*.[6] Vismann, writing within the tradition of German media theory, particularly theories of cultural techniques, is interested in ontology and the development of concepts. At its best her book uses legal files as an example to think about new ways of conceiving history and philosophy through a media and theoretical lens. This book might be described as complementary to Vismann's, expanding the scope of the terrain to think more explicitly about power and epistemology and the way the ontological grids that Vismann highlights get activated in certain institutional settings to establish social dynamics and relationships.

I do not want to understate how important Vismann's work and other critiques and debates about the status of technologies as objects are to how I think about the filing cabinet. But the history of the filing cabinet in this book demands a different approach, one where power is not softened or considered in a register of abstraction that makes it irrelevant. This history of the filing cabinet centers gender and capitalism to understand the physicality of the filing cabinet and to grant it "soft agency." To approach the filing cabinet from this perspective is to grasp the significance of ideas of efficiency and femininity for the development of information as a discrete unit, a thing that can be extracted, circulated, and exchanged; information as something you can have at your fingertips starts with the fingers of women. To center these power dynamics is to write a vertical history of information.

Introduction

THE EFFICIENT WORK OF PAPERWORK

The vertical filing cabinet: a rectilinear stack of four drawers, usually made of metal. With suitable understatement, one design historian has noted that "manufacturers did not address the subject of style with regard to filing units."[1] The lack of style figures into the filing cabinet's seeming banality. It is not considered inventive or original; it is simply there, especially in twentieth-century office spaces. Its ubiquity, along with its absence of style, perhaps paradoxically also contributes to the easy acceptance of the filing cabinet's presence, which rarely causes comment. In films and television shows produced through the twentieth and early twenty-first centuries, the appearance of a filing cabinet defines a space as an office but rarely draws attention to the work it does in that office. One or more filing cabinets line the walls of offices of newspapers, advertising agencies, and other organizations, as well as those of doctors, attorneys, private eyes, and police inspectors. However, occasionally the neatness or disorder of a filing cabinet is intended to give viewers an insight into the mental state and work practices of the office's occupant. The filing cabinet also sometimes performs an important supporting role in dystopian critiques of bureaucracy.

A book about the filing cabinet is not going to fall prey to the allure of ubiquity and the suggestion that the filing cabinet is ordinary. Instead, the filing cabinet's pervasive appearance in films and television shows, especially as background in offices where it

1

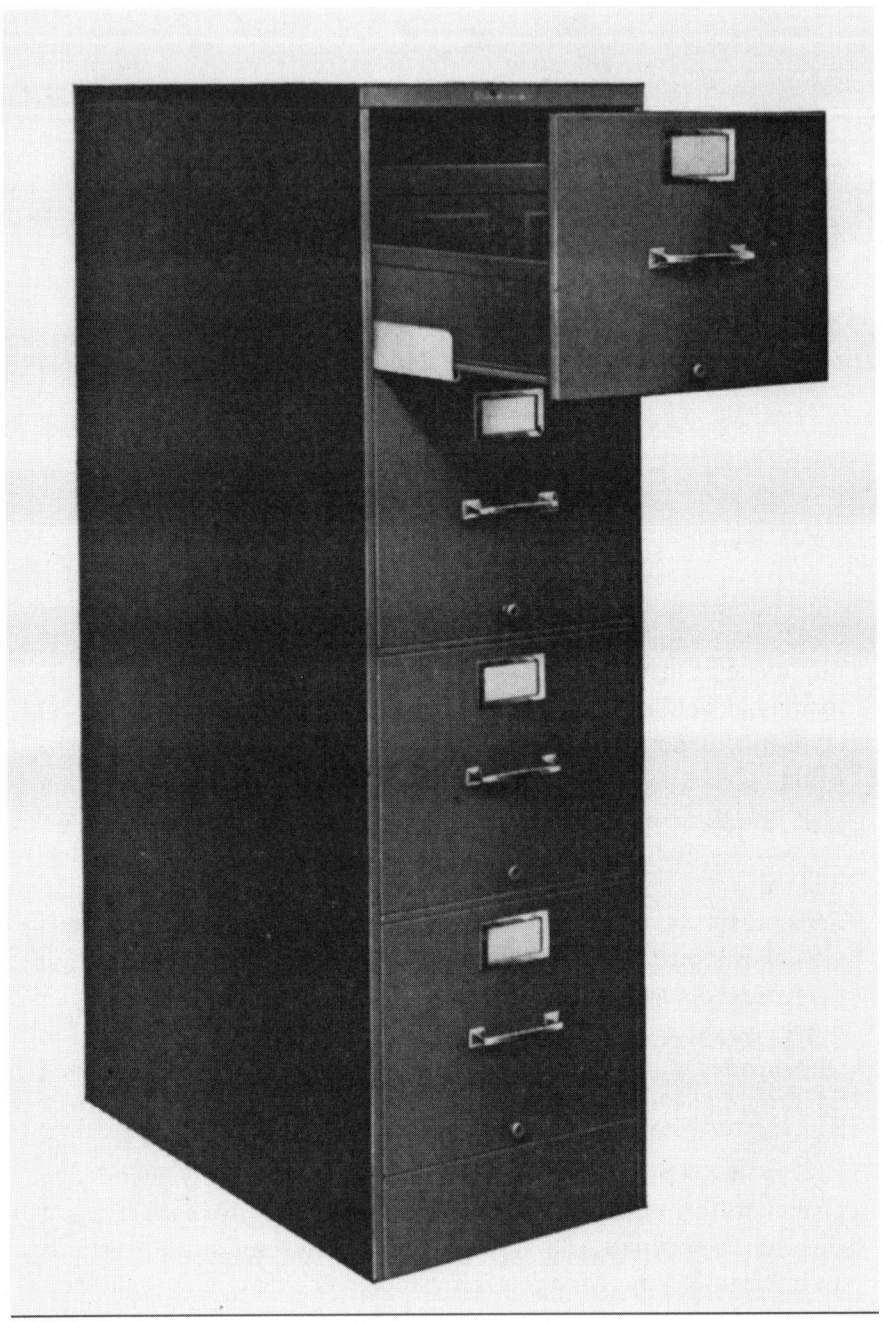

Figure I.1. This catalog page for Globe Art filing cabinets illustrates the most common four-drawer cabinet, with drawers sized to hold letters and other papers of similar dimensions. Globe-Wernicke, *Steel Filing Cabinets* (1931), Trade Catalog Collection, Hagley Museum and Library, Wilmington, Delaware.

contains files with information about people, lays the groundwork for my argument for its importance. A filing cabinet does not just store paper; it stores information. If the modern world depends on information, the filing cabinet needs to be recognized as critical to the expansion of modernity. In recent years academic scholarship has given attention to filing systems used for the storage and retrieval of information critical to government and capitalism, particularly information about people—case files, identification photographs, credit reports—and more generally the development of capitalism in its corporate mode.[2] However, the focus on filing systems ignores the place where files are stored; only in a handful of exceptions from the fields of design and library studies have analyses of filing focused on the filing cabinet, albeit with investments different from those in this book.[3]

Could capitalism, surveillance, and governance have developed in the twentieth century without filing cabinets? Of course, but only if there had been another way to store and circulate paper efficiently; if that had been the case, that technology would be the object of this book. This book focuses on the filing cabinet because it was critical to the infrastructure of twentieth-century government and capitalism; it shares with most infrastructure the fact that, embedded as it is in everyday practice, it is rarely considered worthy of comment, and the labor associated with it is minimized or ignored.[4]

As *the* response to the problem of storing paper and information, the filing cabinet emphasizes distinctive material affordances and economic and cultural priorities. These include efficiency, exploitation of gendered labor, anxiety over information loss, and what I call *granular certainty,* the drive to break more and more of life and its everyday routines into discrete, observable, and manageable parts. This drive is evident in the immediate context in which the filing cabinet emerged: the project of scientific management, the manufacture of interchangeable parts, occupational specialization and professionalization, and the logic of bureaucracy as described by Max Weber, who emphasized clearly defined domains of authority and the benefits of interchangeable staff while briefly noting the importance of the file.[5] Significantly, these affordances and power dynamics continue to exist in the "files," "folders," and "tabs" (and "desktops") people use today to interact with digital information and data.

In this book, I examine the affordances of the filing cabinet as an information technology by explicitly considering the relationship between power and epistemology.[6] Specifically, I argue that distinctive concepts of storage, filing, information, and efficiency get activated in certain institutional settings to establish social dynamics and relationships, especially those of gender and labor. Two overlapping sets of arguments are critical to this analysis: first, the filing cabinet allows us to understand an important moment in the genealogy of information's ascendancy; and second, the filing cabinet is an example of the material history of efficiency.

Briefly, the vertical filing cabinet brings to the foreground a commitment to particularization that shaped the use and conception of information at the beginning of the twentieth century. The filing cabinet contributed to the development of a popular nontechnical understanding of information as something discrete and particular. Critically, it illustrates the moment in which information gained an identity separate from knowledge, an instrumental identity critical to its accessibility. In its separation from knowledge, information was granted authority based on a set of ideas, practices, and institutional supports that limited interpretation; in contrast, a person, a "knower," underwrites the authority of knowledge.[7] This moment in the genealogy of information's ascendancy is tied to broader changes in society driven by the emergence of corporate capitalism and the late nineteenth-century articulation of efficiency that made "saving time" one of the defining problems of the twentieth century.

An investment in particularization produced "system" as a technique to achieve efficiency. Specifically, the vertical filing cabinet shows how system—as it was conceptualized through scientific management, manifested through a vertical predisposition (cabinet and skyscraper), and enacted through a disciplined female body (employed in a feminized clerical job)—made paper and information functional in the twentieth-century office. As a specific example within the material history of efficiency, the filing cabinet illustrates how standardization, rationalization, precision, and speed became skills that workers had to master.

From these perspectives, the discussion in this book gives historical context to current debates over big data. Nick Couldry summarizes big data as "a general project of social re-construction (and, in a sense, reductionism) . . . driven by huge corporate and governmental

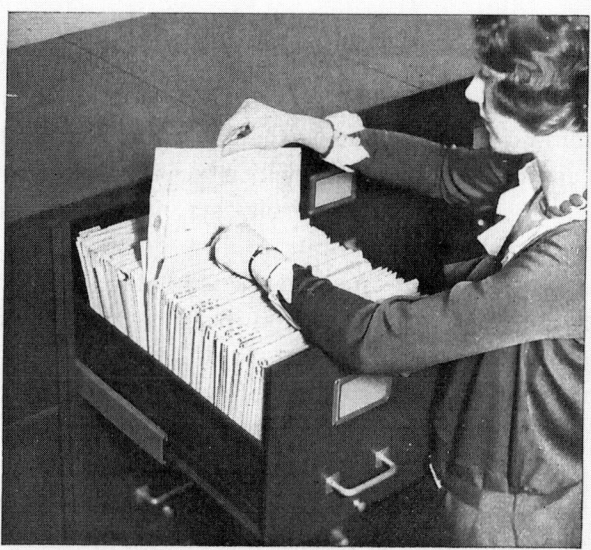

Are letters on your desk in stop-watch time?

D ON'T blame your file clerk when she's slow in bringing you something you ask for—until you are sure the fault doesn't lie with the file.

What kind of filing equipment has she got to work with? Anything as efficient as the Art Metal 6700 File shown above?

There's a file that's planned for modern business! You get every inch of filing space you pay for . . . patented ball-bearing roller suspensions make drawers accessible to full capacity. Make them slide smoothly, too . . . they literally *coast* in and out.

The 6700 File has a special *positive lock* compressor that keeps papers smooth and firmly in place—yet a slight pressure of thumb and finger releases it. That compressor wastes no space, either.

And this Art Metal File will last a lifetime. It's framed of electrically welded steel . . . with cross bars at each drawer to make the whole cabinet rigid.

There are nine different 6700 styles. And they are only one group of the 81 types of Art Metal vertical files that cover every possible filing need — just as the complete Art Metal line covers every office equipment requirement. Every Art Metal product is of lasting, warp-proof steel . . . finished in fine wood graining or rich olive green.

We shall be glad to furnish information on office equipment for your type of business. Or, if you need more equipment for your present office, just check below the kind you want and we will forward a catalogue.

Art Metal Construction Co.,
Jamestown, N. Y.

☐ Fire Safes ☐ Horizontal Sectional Files
☐ Desks ☐ Upright Unit Files
☐ Plan Files ☐ Counter Height Files
☐ Shelving ☐ Postindex Visible Files

Steel Office Equipment, Safes and Files

Figure I.2. This advertisement for Art Metal filing cabinets illustrates the overt attempts to market the filing cabinet as efficient. The advertisement references the stopwatch to connect the filing cabinet to the idea of efficient work championed by the project of scientific management. Courtesy of the Fenton Historical Society, Jamestown, New York.

resources, and focussed on the re-gearing of social order so as to better serve capital's drive to generate economic value from data."[8] Today, as in the early twentieth century, a specific conception of instrumental knowledge privileges the practical (and the socially actionable) as the criteria through which social knowledge is valued. To be clear, this book is not about big data, nor does it offer a direct origin story for the contemporary project of datafication except to show that similar debates have happened before. Rather the early twentieth-century version (the identification, via efficiency, of information as an instrumental mode of knowledge) highlights a critical moment in the ascendancy of information as a defining aspect of contemporary society. The coming to power of information, "the right of a certain produced sense of information to claim our future," is a necessary condition for the emergence of big data to challenge the conception of social knowledge and to shape everyday life.[9]

The Beginnings of the Filing Cabinet

Invented in the United States in the 1890s, the vertical filing cabinet quickly became a fixture in offices in North America and abroad. It spread globally because it provided a way to store large amounts of paper so that individual sheets could be retrieved easily. The method of using drawers for storing paper on its long edge was significant because loose papers cannot stand upright on their own. Put another way, the filing cabinet provided the technological support to enable loose paper to stand on its edge so that more paper could be stored in less space but still be accessed with minimal difficulty. It allowed loose papers to do the work of paperwork.

How does a filing cabinet do this work? According to patents, a filing cabinet functions because its manufacturers drew on techniques and practices from cabinetry and metalwork in new and useful ways. In a patent, a filing cabinet is a collection of steel plates, rollers, slides, walls, ball bearings, rods, flanges, corner posts, channels, grooves, locks, tops, bottoms, sides, arms, legs, and tongues. All of these were used to create a cabinet that would help a drawer open and close when full of paper that could weigh upward of seventy-five pounds. The thousands of sheets of paper that manufacturers claimed could fit in a file drawer were organized using guide cards and manila folders, both accented with tabs. These helped paper stand vertically on its edge, but, more important, they made visible the organization of the papers. Early guides to filing quickly identified the key principle of vertical filing as "the filing of papers on edge, behind guides, bringing together all papers, to, from, or about one correspondent or subject."[10] Storing papers on their edges in this

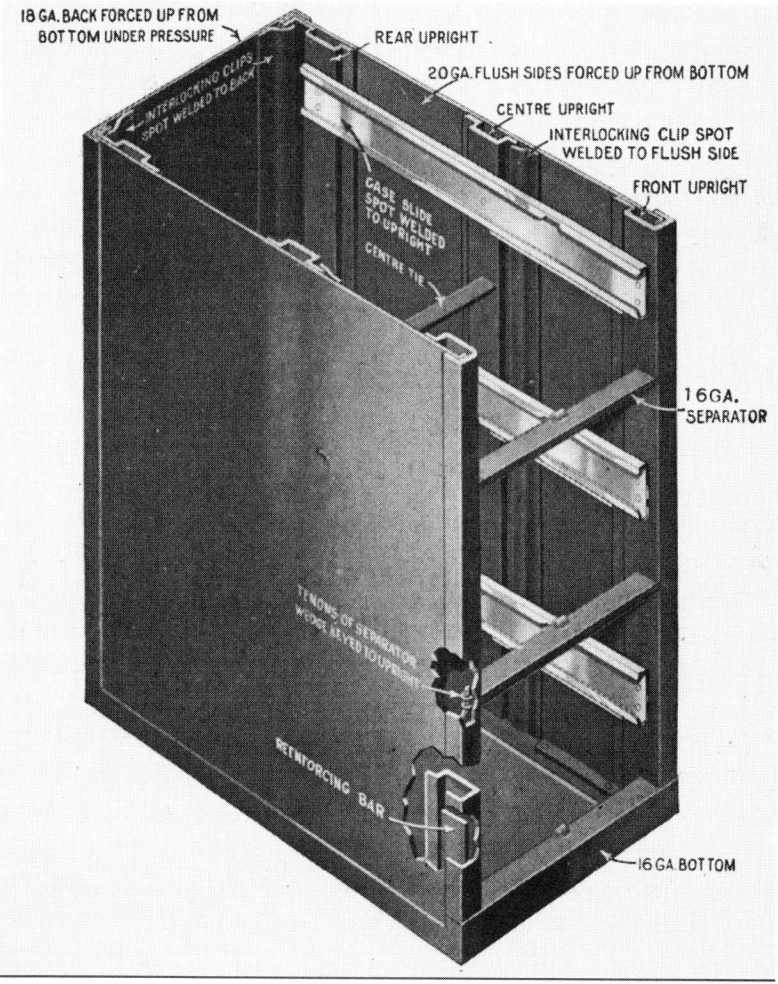

Figure I.3. Filing equipment companies took great pains to show the range of parts and techniques that went into the manufacture of a filing cabinet, as this page from a Library Bureau catalog shows. Author collection.

way made particular pieces of paper easy to locate and easy to access: "The flat file permits the use of but one hand, while with the vertical file both hands are used, thus increasing speed. That is, *papers filed vertically are accessible, compact, and sanitary*" (advocates believed the last of these characteristics was critical to the health of an efficient worker).[11]

Contemporaneous accounts of the filing cabinet tended to link it to the spirit and ideas used to define the period as the high point of modernization and industry. As a response to the idea that modern business needed to do things "quickly, surely, and yet safely," the filing cabinet was offered as evidence of the contemporary "Spirit of Speed." One company, in an attempt to raise the status of the filing cabinet, linked it to "the telegraph, the telephone, the adding machine, the dictating machine."[12] The filing cabinet was a "machine" that encouraged speed and productivity. By identifying the filing cabinet as "modern," celebrants presented its invention as inevitable. Most of the filing cabinet's component parts were not innovative. Boxes, drawers, and cabinets had been used to store papers for hundreds of years previously.[13] Apparently all that was needed were ideas of efficiency and system: a fixation on speed and productivity and a belief that to achieve both, administrative procedures needed to be rigorously standardized.

Given that the vertical filing cabinet was a product of the times, it is not surprising that it had at least two inventors—likely along with others who never made it into the historical record. The current accepted version attributes the invention to the Library Bureau. Melvil Dewey, inventor of the decimal system of library classification, founded the Boston-based company in 1876, although by the 1890s he no longer had any day-to-day involvement with the company.[14] The Library Bureau would later proudly claim the invention, but the critical steps in the development of the filing cabinet took place outside the company. The secretary of a Buffalo, New York, charity organization provided the initial impetus for construction of a vertical filing cabinet. The man, identified as Dr. Rosenau, inspired by the use of cabinets to store index cards on their edges, sought a bigger container for papers. In 1892, he took his idea to the Library Bureau's Chicago office, which built a prototype. Prompted by a "business systematizer," the Chicago office made cabinets for business correspondence, albeit for the note-sized paper used for a lot of such correspondence. However, the company's Boston and New York offices, in a short-term challenge to the narrative of inevitability, saw little value in the invention, so sales of filing cabinets, few in number, were limited to the Midwest.[15]

A competing origin story may explain why the Library Bureau decided to promote the vertical filing cabinet more aggressively in the late 1890s. In this story, Edwin Seibels, an insurance salesman, built a prototype in 1898. A year later, a Cincinnati-based company, Globe-Wernicke, produced five vertical filing cabinets, one of which is in the collection of the Smithsonian Institution's National Museum of American History. Seibels was denied a patent for his invention, not on the grounds that another patent had been granted, but on the grounds that his application described a system that was an idea, not a device. Counter to the arguments of this book, the patent office downplayed the importance of the cabinet, arguing that a box of any size could be made that would not infringe the patent. Reflecting on this decision later in life, Seibels thought he could have successfully patented his use of envelopes and guides as a system.[16]

However, it is unlikely that either Seibels's failed patent or his neglected patent would have made him any money. The turn of the twentieth century saw the rapid growth of an office equipment industry in which dozens of companies manufactured practically identical products with little respect for the hundreds of patents issued for products and parts.[17] To establish its uniqueness, this industry explicitly called its products "equipment," "appliances," and "machines," not furniture.[18] The companies that featured prominently in the development of the filing cabinet were located in the Midwest and Northeast United States. Michigan was home to a number of companies, many of which began in Grand Rapids or close to it; the city had established itself as a center for the manufacture of home furniture. Michigan-based companies included Wernicke, Fred Macey, Shaw-Walker, and Metal Office (which became Steelcase, the only original office equipment company to survive into the twenty-first century). Ohio was home to Globe (which would merge with Wernicke), General Fireproofing, and Safe-Cabinet. In upstate New York, Art Metal, Yawman and Erbe, and Rand were important companies, along with the factories of the Boston-based Library Bureau. In step with the times, the 1920s saw several mergers that reduced the number of office equipment companies, the most significant of which resulted in the creation of Remington Rand. The *Wall Street Journal,* invoking a corporation that had gained notoriety for its purchase of rival companies, called Remington Rand the "General Motors of office appliance companies"; the same article also noted that most Wall Street denizens likely knew nothing about the companies involved due to the lowly status of the office equipment industry.[19]

Despite its purportedly humble status, the industry was successful in making its products indispensable to offices; in so doing, it helped to constitute the office as a "modern" work space. The office that a filing cabinet found itself in was a different space from its nineteenth-century equivalent. The breakdown of the work of a general clerk into specialized tasks underwrote that change; the office equipment industry provided products to facilitate that

Figure I.4. This promotional postcard circulated by the Boston-based Library Bureau shows the company's filing cabinet factory in upstate New York. The factories for filing equipment manufacturers grew in size as the demand for filing cabinets increased in the early twentieth century. Author collection.

specialization. Women operated this new office equipment, with their work illustrating the gendered labor critical to the twentieth-century project of efficiency.

Michigan-based Shaw-Walker's "Built Like a Skyscraper" advertising campaign highlights the gendered understandings of work and information the filing cabinet made pervasive in the early twentieth century.[20] Its trademarked image linked masculinity and modernity: a drawing of a man in a suit jumping into an open filing cabinet drawer, with a sketch of New York City's Woolworth Building, then the tallest skyscraper in the world, in the background. This image left visible the steel frame at the top of the building: to

Figure I.5. This image from a Shaw-Walker catalog is an example of early twentieth-century filing cabinet advertising, which often engaged male bodies to promote the strength of the cabinets. The intended message was that if a cabinet could support a man sitting on or jumping into the drawers, then a user could expect the cabinet and its drawers to function when full of paper. Shaw-Walker, *Built Like a Skyscraper* (1927), Trade Catalog Collection, Hagley Museum and Library, Wilmington, Delaware.

build a filing cabinet "like a skyscraper" was to construct a steel skeleton strong enough to support the weight of the drawers.

Shaw-Walker's advertisements constructed a series of physical encounters between male and female bodies and the company's filing cabinets to illustrate different aspects of the "essentials of office equipment": strength, rigidity, easy operation, noiselessness, economy of floor space, maximum capacity, and good design. In addition to jumping into open drawers, men were depicted lifting their bodies off the ground and hanging from open drawers (what the catalog called "handstands"). The latter images were used to signify the "rigidity" of the drawers as opposed to the "strength" illustrated by men jumping into them (other companies made similar points, using photographs of men sitting on or standing in the open drawers of their filing cabinets).

The "Built Like a Skyscraper" campaign was not subtle. It does not take much for twenty-first century scholars armed with theories of gender representation to argue that the brief exercise routine seen in the advertisements reflected the anxiety men felt about the arrival of women clerical workers in offices, particularly those men who also worked as clerks. The phallic skyscraper, the unsheathed tip of the Woolworth Building, the rigid and erect athletic male body—all sought to make explicit the masculinity of the men who worked in offices; such masculinity was not to be questioned, including that of the men at higher levels in the office hierarchy, who "thought" their way through the day. The image the campaign used to demonstrate "easy operation" of the filing cabinet illustrated the gendered division between manual work and mental work. In contrast to the conflation of strength and masculinity, this image used a female body to show how easy it was to open and close a full file drawer weighing seventy-five pounds: a drawing ("based on an actual photograph") showed a young girl opening a fully loaded file drawer by pulling on a silk thread. The decision to use a young girl rather than an adult female file clerk was deliberate: it clarified that anyone could operate a filing cabinet. Therefore, if anyone could file—if filing required only the strength of a girl—then only women should file, because men could do other work that women could not do.

In addition to the opening and closing of the drawers, the use of

Figure I.6. These illustrations from a Shaw-Walker campaign apply dominant ideas about gender, strength, and work to the selling of filing cabinets. Such images helped to naturalize the changes that brought women into offices as workers. Shaw-Walker, *Built Like a Skyscraper* (1927), Trade Catalog Collection, Hagley Museum and Library, Wilmington, Delaware.

the filing cabinet's interior arrangement of tabs to locate specific pieces of paper, to find information, became associated with women: to handle a tabbed manila folder was to experience an encounter with information shaped by gender. A woman who filed needed to handle papers but not to understand their subject matter. The filing cabinet was marketed as doing the thinking involved in storing and retrieving information. Loose paper was stored on its edge in folders with tabs that were organized according to classification systems. In this way a filing cabinet organized the location of information; it was promoted as "remembering" where papers were so that filing clerks would not need to. The gendered division of the modern office assigned women to assist men; in contrast, men read the files, doing work understood to require thought.

Thinking about the Filing Cabinet

Filing is a particular mode of work that involves an encounter with a distinctly modern and instrumental conception of information: the women who filed only needed to grasp files, they did not need to comprehend their contents. To that end, it is important to note that through the first half century of its existence, in both American and British English the object of this book was called a *filing cabinet,* not a *file cabinet.* I use the former in this book not only for historical accuracy but also because its unambiguous emphasis on the action, not the object, foregrounds the importance of the filing cabinet as a site of labor.

To think about the filing cabinet from the perspective of filing is to argue that "ways of doing" are important because they "engender systems of knowing and modes of social organization."[21] Shannon Mattern's concept of "intellectual furnishings" succinctly captures the stakes involved in this research. Mattern uses the concept to connect the histories of technologies such as bookshelves, desks, card catalogs, and server racks. She argues: "I recognize these furnishings as much more than utilitarian equipment; instead, they scaffold our media technologies in particular ways, inform the way human bodies relate to those media in particular ways, and embody knowledge in particular ways. They render complex intellectual and political ideas material and empirical."[22]

To assert that technologies "embody knowledge" and "inform"

the way bodies interact with them is to make an argument about the materiality of technology.[23] Materiality offers the potential to move away from debates about technological determinism, a label applied to arguments that seem to suggest technology is the singular cause of social change. In contrast to the limitations a critique of technological determinism introduces, materialist theories of technology at their most useful offer a more nuanced understanding of the relationship between technology and society.[24] Such theories become useful when they explicitly acknowledge the intersection of power dynamics and technology. This approach argues that materialism should "not dismiss questions of human agency, of meaning, or of interpretation, but must stress that the physicality of media plays a part in defining the limits and possibilities in which all of these come to matter and make sense."[25]

This understanding of materialism does not look at an object

Figure I.7. This cover for a brochure advertising Yawman and Erbe's 800 Series shows how women were represented as the ones who filed in offices from the moment the filing cabinet emerged. A woman filing always stood tall, her height matching the verticality of the cabinet to illustrate ease of access to files, even in the top drawer. Courtesy of the Smithsonian Libraries, Washington, D.C.

like the filing cabinet as inert or frozen. It thinks about the object in operation, as something that is activated. It considers the action of filing, the fingering of tabs and folders, the moving of drawers on slides. A filing cabinet changes a person and an institution through its operations. Prior to the invention of the filing cabinet a person did not file by handling tabbed manila folders. What did it mean that women were now doing that? What did it mean that institutions started to store their records in rectilinear cabinets as loose, unbound pieces of paper? Filing cabinets were many people's first encounter with a conception of information as discrete units. This interaction needs to be understood as a way in which people learned to be efficient. As such, this acknowledges what makes this approach to the study of technology and culture valuable: it takes broad, abstract, and (often) imprecise notions and ideas and observes them through an object.

Therefore, in this book I examine the filing cabinet as an object that is activated in particular ways, an object that does some things and not others. It is understood to have emerged as a solution to a set of problems. However, as a response to those problems it generated a set of processes that affected thought and action. This is the framework through which I position the filing cabinet in a material history of efficiency and use it to explain a particular moment in the genealogy of the ascendancy of information.

Information and Efficiency

My aim in this book is not to make the filing cabinet the origin of change; I do *not* claim that the filing cabinet invented a modern conception of information or that it invented efficiency. I do argue, however, that the filing cabinet provides an important way to understand how information would become the defining aspect of an "age" and "society"; it gives historical context to an important phase in the ascendancy of information.

No one in the first half of the twentieth century claimed to be living in an "information age"; however, some people increasingly recognized the need to name something different and distinct from knowledge.[26] Sometimes writers identified it by appending a clarifying adjective, using terms such as *classified knowledge*; others named it *data* or simply *information*. The interchangeability of these labels signaled the novelty of this conception of knowledge; in the early twentieth century it was clearly more important to identify the existence of an instrumental form of knowledge and to link it to ideas of efficiency and productivity than to give it a consistent name. The connection to these economic values differentiates this investment in the specific and particular from earlier moments in the history of knowledge, for example, the development and use of reference tools by early modern scholars.[27]

In the American business world, *information* became a preferred label. The magazine *Machinery* informed its readers in 1912 that useful information was "carefully and systematically collected" and then "classified and digested." Created through procedure, such information could be presented as superior to "individual judgment." This process took knowledge and made it "accurate information," easily understandable by anyone and "instantly available whenever a problem is presented to management."[28] This "information" existed in the world with its meaning purportedly transparent to all. Although this is not the information of mid-twentieth-century information theory, it was most definitely a "thing" primarily understood as a noun, not a verb, having lost its earlier association with the process of being informed.[29] The promotional literature for the filing cabinet explained the value of "information" by extolling the necessity of efficiency and productivity; information ensured precision, accuracy, and speed. Picked up across a range of sites, the commitment to particularization moved the still somewhat novel ideas of rationality and management beyond industry.

I offer the concept of granular certainty to explore the critical connection between efficiency and information. At the turn of the twentieth century, *efficiency* became the "catchall word for cost-effectiveness and control of the work process."[30] Efficiency quickly became a nimble set of ideas. However, regardless of its deployment, efficiency in part depended on the belief that breaking something down into smaller pieces made it easier to apprehend, understand, and control; to create something small was to guarantee certainty. Notably, that "something" tended to be labor, reconfigured by Frederick Winslow Taylor's project of scientific management as the interchangeability of standardized parts—a set of ideas that make explicit the role of engineers in the origin of the profession of management.[31] This version of efficiency introduced a set of temporal concerns that centered on reducing the time it took to complete a task as well as a focus on the future through an emphasis on knowing what needed to be produced and when. Efficiency made planning the key to economic growth; planning and control needed information.[32]

However, the analytic value of granular certainty is that it moves the focus away from understanding the need for information to understanding how information was conceptualized, constituted, and organized in practice. This shift emphasizes the overlap between the importance of efficiency's embrace of standardization and the particular and a conception of information as something discrete. The vertical filing cabinet and tabbed manila folder emerged from this overlap between efficiency and information.

Tabbed files emphasized information as something that existed in discrete units, in bits, that could be easily located and extracted depending on the needs of a specific task. Granular certainty highlights the specific moment when the classification of information became a temporal and spatial problem—that is, the moment when classification became a problem of efficiency, a problem of labor, and when storage became a problem of rapid retrieval. In filing, granular certainty manifested a belief that increasing the number of subdivisions would enhance the probability that specific papers would be located and information found in a timely fashion.

Secretary of State Elihu Root's frustration with the result of his request for a handful of letters suggests the way in which people valued the filing cabinet at the turn of the twentieth century—as I have argued in the Preface, the filing cabinet offered a way to store sheets of paper so they could be easily accessed as individual units. The arrival of several bound volumes on Root's desk emphasized the ongoing importance of the book to the late nineteenth-century office. The book had long satisfied the storage needs of offices, particularly for correspondence and accounts. When Root received a letter in a book, it was a piece of something. The letter was physically tied to the larger context of the actions of a particular consul, whose complete correspondence with the State Department was bound in the volume. The letter was part of a larger whole because of the serial relationship created by the binding; as a page, the letter was "both unique and serially adjacent to what followed, which was also structurally similar."[33]

FILES
... THE TREASURE CHESTS OF BUSINESS

Figure I.8. This image from Yawman and Erbe's 1931 catalog featuring steel filing cabinets illustrates one technique manufacturers used to promote cabinets as efficient. In a business world where information had become valuable to planning, a filing cabinet made information accessible through the use of tabs. Yawman and Erbe, *Steel Filing Cabinets That Serve Discriminating Investors* (1931), Trade Catalog Collection, Hagley Museum and Library, Wilmington, Delaware.

In contrast, the filing cabinet prioritized another relationship between paper and knowledge and a different set of expectations about memory and forgetting. In the absence of the larger whole of a book, a letter extracted from a file drawer (and often a folder) offered a new mode of reading. It remained the same letter, but the reader encountered it through a heightened degree of granularity, where it existed as a particular object. No longer a page, the sheet of paper was stored and retrieved as a discrete object in a fixed place, located based on a specific aspect of its content. It was still part of something larger, the folder and the filing cabinet, but as a discrete object the letter was experienced in its particularity. A sheet of paper, as opposed to a page in a bound volume, offered in its material form a different degree of granularity that supported an informational mode of reading.

A number of scholars have argued that a distinctly modern conception of information became comprehensible in part through a connection to the properties of paper documents, "separate and separable, bounded and distinct."[34] Loose paper gave a material existence to information as a thing that could be detached and repositioned, reordered, and recombined. This increasingly pervasive investment in particularity is connected to the "enthusiasm for classification as a technology of search and retrieval" that Alex Csiszar argues became important in the last decades of the nineteenth century.[35] An instrumental use of classification occurred across multiple sites, including philosophy, science, law, and librarianship. Discrete knowledge, recognizable as data, became critical to natural history, paleontology, and census work in the nineteenth century. Staffan Müller-Wille contends that these developments turned "species and other taxa into objects that could be counted and whose numbers mattered."[36] The universalizing projects of "documentalists" such as Belgian Paul Otlet sought to abstract information from all published books. In Otlet's words (as translated into English), the aim was that "each item of information has its own identity."[37] Similarly, in the early twentieth century, the work of company libraries centered on abstraction, or the extraction of what librarians identified as "information" from whole documents and books; one proponent called his company library a "library of fragments."[38]

In this historical period, filing technology provided a conceptual gateway for understanding information as a thing that could

The Old Style Flat System.

Reference to correspondence filed in the old style Flat Letter Files, in Copy books and in cumbersome Transfer Cases involves *time, trouble, annoyance and worry,*—the correspondence is *always* scattered,—letters received and copies of answers are *never* together, indexing is *seldom* kept up to date and correspondence from different sources is *almost always* mixed.

Flat Letter Files.

In the old style Flat Letter File, correspondence is filed flat in drawers, each drawer being about $3\frac{1}{8}$ inches deep, as illustrated and described on page 44. The "Macey" Flat Letter File possesses points of marked superiority over files of similar nature, in construction, selection of material, and in mechanical features. It is the *best file manufactured of its kind* and is recommended to all desiring a letter file of this style, but for quickness of operation and other advantages mentioned page 36, it is surpassed by the Vertical File.

Fig. 52.

Fig. 53.

Figure I.9. This page from a 1901 Fred Macey catalog shows how promotional materials for filing cabinets overplayed the disorganization of storing paper in bound volumes to emphasize the value of storing loose paper in folders. Courtesy of the Smithsonian Libraries, Washington, D.C.

be standardized, atomized, and stripped of context—information as a universal and impersonal quantity. While this conception did not begin or end with the filing cabinet, the file became a common way of making this information comprehensible, as it continues to do in the present with the information and data encountered through digital technology. The file's ongoing role in the metaphorical cluster used to mediate people's interaction with digital information and data, along with its relation to the filing cabinet, is why this book prioritizes unbound paper and the file folder over index cards and punch cards as objects through which to examine the emergence of modern information.

I make no claim to originality in noting the connection between paper files and computer files. The idea was there in the design of early personal computers. In 1984, Apple intended that the icons on the screen of its Macintosh computer would make a user's initial encounter with the technology more comprehensible by making "the manipulation of information in the system analogous to the manipulation of physical objects on a desktop"; this extended the "desktop metaphor" that computer scientists had been using since the 1970s.[39] Scholarship in media studies has noted and critiqued the connection between paper files and computer files. One example appears in the conclusion to Cornelia Vismann's media archaeology of legal files, a version that Wolfgang Ernst uses in a critique of digital archives.[40] Ernst argues that the prolonged life of the "archival spatial order" hinders the move to temporal organization inherent in the continual processing of data and memory in digital media.[41]

However, it is important to note that the "file-like icons" Ernst alludes to in his argument are not the files of "old European times and secretaries and offices" that he invokes from Vismann's arguments.[42] The icons take the form of manila folders, usually tabbed manila folders. These kinds of files, along with the vertical filing cabinets that accompanied their debut on the screens of desktop computers, are the products of a very different historical moment. They are the files of early twentieth-century American times and file clerks and offices.[43]

My intention in distinguishing between files from different historical periods is not to make a pedantic claim to historical accuracy. Rather, by acknowledging the historical specificity of different conceptions of the file, I aim to bring to the foreground the power dynamics that characterize the organizing and ordering of information, and to recognize the significance of shifts in how information is conceptualized, constituted, and organized. It is critical to note that the concepts people use in the twenty-first century to comprehend data and information originated in a particular moment in the history of capitalism, a moment that led to a gendered, raced, and classed understanding of efficiency becoming pervasive.

Collectively, the chapters that follow offer a series of overlapping arguments about information and efficiency. These arguments are explored in two parts, the first of which focuses on the cabinet as a physical structure. The second part of the book addresses how that structure directly shaped people's interactions with paper and information; it is concerned with the act of filing rather than with filing systems.

Part I examines the material properties of the filing cabinet, its capacity as a storage container. To think about storage is to think about the principles that organize the objects that are stored. According to Lewis Mumford, principles of storage arise in response to a range of concerns, including accumulation, continuity, enclosure, and protection.[44] Mumford argues that one of the most significant developments in the Neolithic period was the use of storage containers, including urns and granaries, which enabled humans to collect grain when it was plentiful and protect it for future use when it was less abundant.[45] To store is to create a supply of something, to accumulate stock or stores to the level of abundance. The values invoked in the cause of storage create a designated place to which a person goes, knowing that certain things will be located there.

Integrity, verticality, and temporality are principles central to the storage of paper in a filing cabinet. These principles link the filing cabinet to efficiency and the late nineteenth-century changes in capitalism, particularly the development of corporate capitalism as a way to manage the increased scale in production afforded by industrialization. While accumulation, continuity, enclosure, and protection are all important to the filing cabinet, it was designed and marketed to foreground easy access to paper. Although storage always involves an awareness of future use, it does not always prioritize the moment in the future when a stored object will be accessed, what is now called retrieval. As a practice of stockpiling or piling, not retrieval, storage prioritizes the allocation of space to store something; it is primarily a spatial practice. In contrast, retrieval is conceived as a temporal practice. The focus is on a process, on facilitating the act of finding something. In short, storage takes up space and retrieval takes up time. Therefore, storage does not automatically invite retrieval.[46]

However, when retrieval becomes an issue, it is not simply a problem of access to the place of storage or the availability of the object. Depending on what is stored and why it is stored, retrieval may be about locating specific items. When accessing grain from a granary, a person is not interested in finding a specific piece of grain. In contrast, a person may be interested in finding a specific letter in a collection of correspondence. When it is necessary to find something particular, the problem of retrieval is usually solved through the organization of space to create a system of order that can be seen, a system that can guide a person to a specific object.

Chapter 1 begins with the oft-commented-on association between the filing cabinet and the skyscraper to foreground the importance of verticality as an organizing principle in the articulation of storage and retrieval in the filing cabinet.[47] Co-opted to the project of efficiency, verticality became a strategy to facilitate productivity by organizing space to make better use of time. To consider verticality is also to locate the skyscraper and the filing cabinet within a heightened awareness of economies of access. This draws attention to the filing cabinet's placement in an office, a space increasingly approached as a productive whole, as a system. The flow of people and paper became critical to how people thought about the office and the design of the spaces occupied by office furniture and workers.

Chapter 2 examines integrity as a critical value in the storage practices of the filing cabinet. The integrity of the filing cabinet references its wholeness as an upright object, a vertical enclosure in which paper is stored. This was linked to the construction of the filing cabinet, especially when steel replaced wood and an object "built like a skyscraper" could be presented as an authentically modern object. Claims to integrity were also important for assuaging anxiety about lost pieces of paper as storage moved away from letter books and bound ledgers to folders and drawers.

Chapter 3 moves inside the filing cabinet to examine the system of order that was built into the structure of the cabinet. The discussion of this "cabinet logic" illustrates how material features supported or constrained the partitioning of paper content into bits of information. This particular cabinet logic was distinct to the late nineteenth century, not because partitions in cabinets were new but because partitions and dividers were being mobilized in cabinets through a late nineteenth-century conception of efficiency. Technologies such as manila folders, guide cards, and tabs made the particular visible, while compressors ensured that papers stood to attention "vertically."

The second part of the book shifts the focus to filing, to how the filing cabinet shaped the acts of inserting and removing manila folders and sheets of paper. In mediating the service work associated with clerks and secretaries, new machines intensified a very specific relationship between worker and information.[48] From this perspective, the arrival of the typewriter mechanized the act of writing, and the introduction of the filing cabinet mechanized the act of remembering. It is important to understand that this experience of information as an object to be processed, stored, and accessed was a historically specific form of "information labor." In this functional and gendered relationship with information, the purpose of labor is neither interpretation nor understanding of the content of information forms.

Filing is an example of a distinct mode of information labor that emerged at the turn of the twentieth century and became critical to the development of corporate capitalism. In the history of the office, information labor highlights the redefinition of office work as machine work centered on the recording and circulation of information on paper. This conception of information labor depended on a feminization of office work used to naturalize the specific deskilling associated with modern clerical work. From the late nineteenth century, the arrival of women in the office to perform the specialized tasks produced by efficiency was naturalized through the association of women with "nimble fingers," given both the kinds of work (such as light manufacturing in textiles) and the leisure activities (such as piano playing) women were expected to engage in.

Chapter 4 uses granular certainty to explore the set of ideas that linked the office and "system." An exploration of efficiency and management shows the importance of information and filing to early twentieth-century scientific management. Although these ideas provided the conceptual space for the filing cabinet, few offices rigorously adhered to scientific management. However, examples of the use of filing cabinets in government, manufacturing, library, educational, and commercial contexts indicate that office equipment provided an indirect but important encounter with ideas of system and efficiency.

Chapter 5 explores how ideas of efficiency represented the filing cabinet as an "automatic memory"—a machine that remembered. With mental work delegated to a "machine," a distinct form of manual work emerged in the early twentieth-century office. This mode of information labor used gender to organize the division between so-called manual work and mental work. Filing, as an example of information labor, illustrates that women's "natural" dexterity coded this as women's work.

Chapter 6 continues to explore information labor by considering the people who filed and the ways filing was taught in schools. When it came to employing women to file, not all women were considered ideal file clerks. In the white, middle-class space of the office, gender intersected with class, race, age, and sexuality. Some white, middle-class women viewed low-level clerical work such as filing as a way to use the self-discipline they associated with efficiency to impart middle-class values to working-class and immigrant women who did filing work, whether the women aspired to those values or not. In a different project, white, middle-class women also attempted (and failed) to make filing a profession equivalent to librarianship.

To conclude the book, chapter 7 moves out of the office and into the home. Few filing cabinets made that move. However, in the early twentieth century, closets and kitchen cabinets developed under the direct influence of the cabinet logic of office storage; speed, efficiency, and an obsession with retrieval marked the emergence of modern planned storage technologies in the home. Therefore, an examination of the arrival of planned storage in the home brings to the foreground the technologies and power dynamics that shaped the specific moment in the genealogy of the ascendancy of information that is the focus of this book. It also illustrates that efficiency gave this moment in information history its specificity.

In this history of the emergence of the filing cabinet, I focus on the changes in storage practices at a critical moment in the history of information. This is important because storage is never a neutral practice. Decisions about what to store and how to store it adhere to social values. This is particularly critical when the object to be stored is knowledge. Storage practices and storage technologies in the office provide an important site for examining a moment in the process through which information became something distinct from knowledge. Information became a name for an instrumental use of knowledge via the importance of the particular in the implementation of efficiency. The vertical filing cabinet offered a way to make sense of this change; it became a way to grasp information, both physically and metaphorically.

Files, tabs, manila folders, and filing cabinets remain critical to the comprehension of data and information and to an understanding of the relationship between technology and gender. To acknowledge the power relations integral to these information technologies is to offer another dimension to critiques about the materiality of information and data. Rather than an infrastructural "tubes and wires" critique, this history makes information and data material by showing that the concepts through which people are still asked to imagine their encounters with information and data (files, folders, tabs) originated in highly gendered understandings of labor and information. It reembodies information by showing that the properties of the information technologies associated with filing are the products of power relations.

The emergence of the filing cabinet makes explicit that an encounter with information naturalized a historically specific set of economic concepts that depended on ideas about class, race, and age as they manifested in gender. This focus gives information a vertical history that is lacking in previous histories of information and files. To call this history vertical is not only a nod to the height of the vertical filing cabinet but also an acknowledgment of vertical power structures and centralization. This history demands that we recognize the importance of who gets to define what information is; control of the conception and use of information is an extension of other forms of control.

PART I
THE CABINET

Chapter 1

VERTICALITY
A Skyscraper for the Office

To be vertical is to be upright, positioned at a right angle to the horizon. The dictionary informs us that to be upright is to be in a vertical position, perpendicular to the ground. The noun *verticality* captures the condition or quality of being vertical or perpendicular. The adverb *vertically* means in a vertical manner, direction, or position. Therefore, to think about the vertical is to consider the organization of space and to think about organization spatially.[1] That is, verticality is a principle, a key attribute, frequently used to organize how objects are stored—a principle critical to the filing cabinet.

Scientific office management guru William Leffingwell understood the innovation and value of the filing cabinet in precisely these terms, arguing that height saved space—that is, it shifted the focus of storage space from the horizontal to the vertical. In his interwar textbook on office management, Leffingwell wrote: "The introduction of the vertical file marked the utilization of a great efficiency principle, namely, that there is always more room for growth in a vertical than in a horizontal direction. When we use the surface only, a much greater area must be covered than when we use the third dimension, height."[2]

Although the word *vertical* in the name was intended to refer to the storage of paper on its long edge, in use *vertical filing cabinet* came to refer to the cabinet itself more than to the specific mode of filing it housed; many people used the name to identify the height

Figure 1.1. The cover of this 1930s Remington Rand catalog illustrates that although *vertical* referred to the system of filing paper on its long edge, it also came to be associated with the height of the cabinet. In both cases verticality illustrated the efficiency of filing cabinets. Courtesy of the Herkimer County Historical Society, New York.

of the cabinet and its rectilinear structure. Leffingwell embraced both uses. For him verticality was critical to the design of the cabinet as a container to support the stacking of drawers and how the filing cabinet stored paper on its long edge in drawers (I discuss the latter in chapter 3). In explaining the efficiency of the cabinet, Leffingwell offered the example of a building: "A one-story building must have one foundation and one roof, but a two- or ten- or twenty-story building requires only the same number of foundations and roofs."[3] Verticality became an "efficiency principle" when people recognized the unused storage space that height offered and created technologies to use it.

Leffingwell's championing of the vertical is an important way to begin to think about the filing cabinet as a cabinet—the purpose of the first part of this book. It emphasizes the most immediate aspect of the physicality of the cabinet. However, to approach the height of the filing cabinet as an example of verticality is also to understand the cabinet's development through a logic critical to capitalism. Verticality locates the filing cabinet within the changes that marked the development of corporate capitalism at the turn of the twentieth century; it makes the filing cabinet available to a material history of efficiency. Efficiency as a way to justify control of the work process involved not only the saving of time but also the saving of space. A somewhat simple example of this is the visual juxtaposition of the filing cabinet and the skyscraper in period advertising; such images promoted verticality as a historically specific remedy to the corporate articulation of spatial scarcity as a problem. Therefore, to better understand the filing cabinet, I decenter it to argue that verticality shaped the filing cabinet, office design, and the skyscraper. The last of these, as the container that housed floors of office space, provided the immediate spaces in which filing cabinets were placed, located amid the concerns about flow and horizontal movement to which verticality responded.[4]

I use verticality to label how height and elevation were arranged and instituted in the early twentieth century—specifically, how height and elevation were aligned with ideas of economics and efficiency. To be clear, verticality as a mode of organization was not an invention of the late nineteenth century, but its co-optation by the business imagination via efficiency was new, and therefore it is

Figure 1.2. This Shaw-Walker catalog cover features both the filing cabinet and the skyscraper. This was a common juxtaposition intended to emphasize not only the verticality of the filing cabinet but also its modernity. Shaw-Walker, *How to File Letters and Cards* (1920), Trade Catalog Collection, Hagley Museum and Library, Wilmington, Delaware.

the focus of this chapter.[5] In this period, the vertical "saved" space by subordinating parts to a greater whole to create order: floors and offices in skyscrapers, and drawers and folders in filing cabinets.[6] It located discrete parts in a structure that involved, or comprised, many levels, often organized as a grid. However, the functionally compartmentalized models that verticality produced to serve the corporate business sector were not solely architectural or design strategies. This vertical predisposition was evident in the wider organization of space, labor, and storage central to the accumulation of capital in its corporate iteration at the turn of the twentieth century (in the second half of the twentieth century, the "network," with its horizontal bias, became more prevalent, and the corporate headquarters that moved to suburban areas saw the architecture of large office buildings turn to horizontal lines).[7]

The vertical bias of capitalism emerged in concert with the rise of the United States as a global economic power in the decades following the Civil War. In this period the development of vertical integration drove capital accumulation (history credits Andrew Carnegie with playing a pioneering role in this process).[8] Other examples from the second half of the nineteenth century emphasize the importance of the vertical to an understanding of order and power, notably hierarchical management and the representational strategy of the corporate ladder. The latter further cemented the modern belief that organization must equal hierarchy and began the process that resulted in the conflation of organization with practices of modern management.[9] C. Wright Mills captures the importance of the concept of verticality to this process in his analysis of hierarchies in mid-twentieth-century office work. For Mills, the hierarchical structure of the office functions like an "enormous file." In addressing the organizational principle of verticality, he argues that "smaller hierarchies fit into larger ones and are interlinked in a dozen ways"; the nesting structure of filing cabinets illustrates the subordination of parts to a whole.[10]

Given the vertical bias of corporate capitalism, it is clear that the decision to store paper vertically was not a case of thinking outside the box (either figuratively or literally). This development occurred within a larger rethinking of storage, one that privileged retrieval and circulation of stored items over their mere containment; saving time and space in relation to labor also involved the organization of movement through space and around objects. Therefore, verticality was a principle used to facilitate movement, which increasingly positioned the retrieval of paper within the collective "flow" of people, paper, and work. The use of verticality as an organizing principle across a variety of sites offers one answer to the question of why at the end of the nineteenth century it finally made sense to store loose paper stacked on its long edge in drawers stacked vertically and housed in a stand-alone cabinet.

Using verticality to solve a problem of paper storage aligned the filing cabinet with its signature location. The filing cabinet emerged at the same time the skyscraper was becoming a cultural symbol that presented capitalism and modernity as distinctly American projects for the twentieth century. In the last quarter of the nineteenth century, business needs, limited space in the developing commercial districts in Chicago and New York City, and new construction techniques all contributed to the emergence of what became known as the skyscraper (early alternative names included skysweeper, cloud scraper, and cloud supporter).[11] By the 1890s, *skyscraper* was the term applied to a building of maximum height, with an interior arranged in floors, one above the other. This particular definition excludes monuments and towers, but it does not limit use of the name to office buildings.[12] However, the skyscraper quickly became associated with the expansion and growth of American business, even though, as one architect subsequently noted, from a practical perspective, the floors in such buildings were used for activities that did not need great height.[13]

In the Shaw-Walker "Built Like a Skyscraper" advertising campaign, however, the visual juxtaposition of the skyscraper and the filing cabinet is not just symbolic. It suggests a shared logic. The images used in the campaign offer the vertical as a framework (in this case a steel skeleton) that could be used to facilitate standardized stacking or to turn a grid into something that could be stacked. Adhering to a central logic of capitalism, verticality *smoothed* height so elevation could be used to accommodate the horizontal. As vertical structures, the skyscraper and the filing cabinet stacked discrete horizontal units (floors containing offices; drawers containing folders and papers). This stacking was intended to serve two main purposes: to reduce the structure's horizontal footprint and to facilitate access to the units. Stacking, as opposed to mere piling, invoked order. Economies of access were critical to the development of these structures. These enclosures of stacked units enabled continuity and consistency through the arrangement of different levels to assist the flow of people and paper.

Skyscrapers

When Shaw-Walker began its "Built Like a Skyscraper" campaign in 1913, the company's decision to use the Woolworth Building to represent a skyscraper was likely a no-brainer. Completed that year, the Woolworth Building took the skyscraper to a new level. At 792 feet, it would be the world's tallest building until the 1,250-foot Empire State Building opened in 1931. While the Woolworth Building was a somewhat unique skyscraper, it clearly represented the vertical predisposition of corporate capitalism and emphasized the skyscraper as an important actor within a material history of efficiency; it was also a manifestation of the ego of F. W. Woolworth, the founder of a highly successful chain of retail stores.

The Woolworth Building was completed in record time thanks to what Gail Fenske describes as "a new level of modern rationalization, efficiency, and financial control over the entire construction process."[14] Fenske also notes that the achievement of the building's record-breaking height depended on "technologically audacious steel-framed engineering."[15] While these facts fit the needs of the Shaw-Walker campaign, the company's decision to use an image of the Woolworth Building also assumed a public that was conversant in the basics of skyscraper construction. For the skyscraper to capture the American imagination, two developments had been critical: first, the production of a specific type of iron (that existed between cast iron and wrought iron), which borrowed the name of steel from precision tools; and second, the inspiration provided by lengthy, but safe, steel spans used to build large bridges.[16] These developments enabled the construction of the skyscraper—a building with a skeleton structure that supported the weight of the walls rather than solid walls designed to support themselves.[17] The sheer volume of steel required to achieve the size of the Woolworth Building was obtainable only because United States Steel had emerged as the world's largest industrial corporation via vertical integration;

Figure 1.3. This postcard was printed as a souvenir of New York City's Woolworth Building, which was the world's tallest building from 1913 until 1931. Office equipment company Shaw-Walker included a drawing of the skyscraper's construction as part of its trademark. Author collection.

F. W. Woolworth himself had developed his chain of five-and-dime stores through vertical integration. Like many corporate heads, Woolworth compared the internal hierarchy of his organization and its perceived strategic efficiencies to a military organization; he was particularly partial to Napoléon Bonaparte. These factors produced a building that, according to Fenske, came from an individual and cultural desire for "a landmark skyscraper that would serve to identify a rapidly maturing, more rigidly hierarchical, still continually expanding, and newly international corporate enterprise."[18]

By the second decade of the twentieth century, the skyscraper had become a symbol for modernity as it incorporated elevation as a "sign of progress, of striving."[19] In this capacity a skyscraper could be seen as both a celebration and a critique of modernity. In advertisements, the view from atop a skyscraper became an "artist's shorthand description for the concept 'modern.'"[20] But as David Nye argues, that view could also be presented as "a businessman's gaze [that] dominated the new man-made landscape."[21] Over the ensuing decades, attitudes toward skyscrapers became even more negative, as the buildings were read not only as "representations of hierarchy and exclusion" but also as "advertisements for ego; thieves of light; . . . and (of course) compensations for having a small penis."[22]

Public celebration of the skyline tended to overshadow concerns about egos and phalluses, but the way in which the verticality of skyscrapers disrupted the horizontal lines of cities was initially harder to ignore. Some American cities, including Boston and Baltimore, restricted the height of buildings. Other cities tried to incorporate the vertical into planning, but the "horizontal city aesthetic" that continued to dominate European cities was often subsumed within a new vertical city.[23] In 1916, New York City passed regulations that accepted the transition from a horizontal to a vertical city but tried to manage access to sunlight so that the public would still be able to view the sky from the street. The resulting setback requirements allowed sunlight to reach the street but also permitted the construction of tall skyscrapers that further emphasized the city's verticality.

However, skyscrapers restricted the experience of height to their exteriors. The division of the interior of a skyscraper into horizontal floors limited the sensation of verticality. This was in contrast to the previously dominant type of vertical building, the church, "where the impact of the interior is of prime importance . . . [because] it conveys the experience and intelligibility of the vertical in a religious context."[24] In contrast, the skyscraper's mode of verticality used height in a more instrumental and pecuniary way (the labeling of the Woolworth Building as a "cathedral of commerce" captures this transition, although the building was given that name for a different reason). The profit-oriented horizontal organization of a skyscraper's interior speaks to the use of the vertical to arrange standardized units, an ordering that depended on uniformity (both externally and internally) to reduce space and facilitate access. As the future architect of the Woolworth Building, Cass Gilbert, put it in 1900, the skyscraper is a "machine that makes the land pay."[25]

Therefore, as a vertical structure, the skyscraper exists to contain repetition; it is a unit that subordinates standardized parts. In early skyscrapers all of the vertically stacked floors were of equal size, with internal divisions repeated. In what we would now label a modular logic, the office became the basic interior organizing unit of the skyscraper. In an article published in 1896, architect Louis Sullivan borrowed the phrase "form ever follows function," twice removed, from biology. For Sullivan, the office was a building block akin to a cell in a body: "The practical horizontal and vertical division or office unit is naturally based on a room of comfortable area and height, and the size of this standard office room as naturally predetermines the standard structural unit, and, approximately, the size of window-openings."[26] Form followed function as the building's facade provided symbolic expression of its interior use through the arrangement of windows.

Frank Lloyd Wright, one of Sullivan's mentees, remembered being in awe when he saw a drawing of an early skyscraper: "There was the very first human expression of a tall steel office building as architecture. It was tall and *consistently* so—a unit where all before had been one cornice building on top of another cornice building. This was a greater achievement than the Papal dome, I believe, because here was utility become beauty by sheer triumph of imaginative vision."[27] Despite this high praise, Wright devoted much of his working life to the horizontal as a response against what he saw as the cultural fetishization of the vertical and the skyscraper in particular. For Wright, verticality quickly grew "stale" as the skyscraper became "a clumsy imitation masonry envelope for a steel structure," a building that was a commercial expedient with "no higher ideal of *unity* than commercial success."[28]

Wright viewed the skyscraper as part of a larger problem in twentieth-century culture, which he called "the culture of congestion." For him, congestion, in the form of restrictions on the ease of movement of people and traffic, was a product of the "tyrannical verticality" the skyscraper represented. Skyscrapers had their place in Wright's vision, but he believed they should be either outside cities, in small towns, or in cities where sufficient space could be maintained between them; he asserted that such settings would give the structure of skyscrapers integrity.[29] Positioned in this way, skyscrapers could exist, but in deference to the horizontal, which Wright associated with space and freedom. In contrast, if skyscrapers as standardized structures were lined up too close to each other, they enacted the dehumanizing potential of modernity. Therefore, as Wright told the viewers of NBC's *A Conversation* in 1953, in his practice he sought to "destroy the box as a building" because the "architecture of freedom and democracy needed something basically better than the box."[30]

Wright was not alone in this critique. For other architects of the time, the skyscrapers of the early twentieth century represented a "conflict between individualism and civic life."[31] From this perspective the city, not individual buildings, needed to be the basic unit of design. The "rampant individualism" of skyscrapers not only restricted the flow of people but also reduced the amount of sunlight that could reach other city buildings and the streets.[32] Hence, a civic aesthetic had to be a horizontal aesthetic.

However, this individualism was critical to the logic of verticality in its early twentieth-century iteration. Skyscrapers articulated a mode of individuality central to capitalism; verticality was an organizational technique deployed for economic ends, not democratic ones. As Leffingwell suggested, the vertical was an "efficiency principle." Verticality was a response to a heightened logic of accumulation in which the needs of production made space scarce. In contrast to a pile, which leads to disorganization and congestion, the vertical was intended to provide structure, to organize through constraint. Through coercion, verticality would force activity and objects into designated spaces; the skyscraper as a vertical structure was a space organized to facilitate the flow of people and information.

If congestion is an accumulation, a crowding together, a heap or pile that leads to the disorganization of regular (healthy) activity, then verticality was intended to avoid it. To the anger of Wright and others, however, only the interior of the skyscraper was subjected to concerns about congestion. That is, verticality managed flow within the skyscraper's structure, but little attention was paid to the effects of the building's centralizing practices on the neighborhood where it was located.

Within the skyscraper, the elevator played a critical role in attempts to "efficiently" organize the rapid circulation of workers and the flow of work. The elevator made the height of a skyscraper accessible by facilitating the flow of people in and out of the building. It helped reduce the congestion the skyscraper created when it brought together large numbers of workers, all of whom needed to access their offices. Prior to the installation of elevators, tall buildings had limited economic value if they rose beyond six stories; six floors' worth of stairs was accepted as the maximum number a person could comfortably climb. After Elisha Otis invented a brake to stop elevators from falling if their cables broke, the usable height of a building increased to twelve stories, the maximum height for a building prior to the introduction of steel skeleton construction techniques.

Even when a tall building had elevators that could be used safely, its profitability was largely determined by the amount of time it took for the elevators to access the higher floors. The Woolworth Building addressed this problem. Its elevators traveled more than twice as fast as those in earlier buildings, and they were carefully managed to allow 14,000 people to go about their work each day.[33] The building used a large number of small elevator cars rather than a small number of large cars, and each car was designated as either express or local, as in a train system. An elevator dispatcher sat in the lobby at a panel with lights that indicated the positions of particular cars; using this information, the dispatcher attempted to keep the cars moving according to a rationalized time schedule.[34] If it ran efficiently, such an elevator system could prevent congestion within the building, benefiting the owners as well as the occupants (however, the system could do nothing about the congestion in the streets surrounding the skyscraper, which Frank Lloyd Wright considered a major problem in the city).[35]

To be efficient and to provide accessibility, a building's verticality had to be legible. The elevator contributed to this by providing each floor with a clear "address," or number. It fragmented the linearity of the vertical into a series of floors represented through electric push-button controls; the late nineteenth-century addition of the doorbell list on smaller buildings further worked to stabilize the category of the floor.[36] Therefore, the doorbell list and the elevator's listing of floors and offices emphasized verticality as a system of order that highlighted the discreteness of (stacked) units to facilitate locatability and access.

Offices

With offices of uniform sizes, the skyscraper as a vertical structure organized space to facilitate movement through repetition. To that end, verticality needs to be understood in relationship to the particular modern articulation of the grid, the cultural technique of using intersecting lines to create cells that locate objects and subjects.[37]

In her provocative work *The Grid Book,* Hannah Higgins recruits bricks, maps, notation, ledgers, type, and boxes to give the grid a long history prior to the nineteenth century. However, as she shows, this history does not reduce the grid's important role in the "visualization of modernity's faith in rational thought and industrial progress."[38] Verticality is one technique that manages the grid, according to ideas of standardization, mass production, and flow; it articulates the grid to the distinctly modern conception of "rationalization by standardization."[39] Wright was appalled by this standardization when it determined the exteriors of skyscrapers, which for him became boxes or envelopes, standardized structures that in turn contained standardized spaces (I discuss the relevance of this to filing cabinets and folders in later chapters).

In her discussion of the rebuilding of Chicago after the fire of 1871, Higgins shows how important the grid was to the manufacturing and distribution process of modern industry. More specifically, she illustrates the importance of what we would now think of as modularity to the modern work environment through the uniformity of office spaces (and apartments) located in tall buildings on streets laid out in a grid. The offices in this grid were "filled with supplies delivered in stackable boxes that could be unloaded into standardized desks with little drawers designed to hold gridlike organizers filled with more standardized supplies."[40] Describing a 1906 photograph of the order-processing room of Chicago-based mail-order company Sears, Roebuck and Company, Higgins notes the "carefully organized grid of identical desks," which contained "uniform-sized clipboards and standardized order forms on which stood legions of clipboards—always to the left—with printed order forms snapped to them." As she argues, "The modern work environment is made of such modules, each module the same as the next, each made up of smaller modules adapted to an incorporated whole."[41]

As suggested in Sullivan's celebration of form, function, and cells, a skyscraper's windows made the grid of an office building visible to the outside world. Windows were critical to offices. While office work did not necessarily need the elevation a skyscraper provided, it did need light. To that end, the height of the building had value as it potentially made more light available to workers, even as early skyscrapers limited the amount of sunlight that made it below to shorter buildings and the street. That is, the need for light affected the size of offices. Windows determined office size, which then determined the size of modular structures within the larger grid, such as desks and filing cabinets.

Given the limited effectiveness of early incandescent lightbulbs, sunlight was the main source of illumination in early twentieth-century office buildings; this remained the case until the arrival of fluorescent bulbs in the 1940s. To enhance natural light, building developers preferred to site skyscrapers with three exposures rather than two; tenants generally desired north-facing offices, with east-facing offices a second choice.[42] Factoring in the reliance on natural light, architects usually designed buildings to be no more than 65 feet wide, with a corridor running through the middle. The distance of 25 to 30 feet from the side of the building to the center was accepted as the ideal "economical depth."[43] Any deeper, and the center of the building would be so dark the landlord would have to charge lower rent. To make the building's center space useful, architects often placed elevators and utilities there. But this "ideal" depth meant that, even on a sunny day, at the 30-foot mark natural light penetrated only enough to produce one-tenth of the minimum acceptable light for offices that experts recommended from the 1950s onward.[44]

Measured between the columns along the exterior wall, an office unit tended to be 18–20 feet wide. With at least two windows in each unit, the design of most buildings allowed for two 9- to 10-foot-wide offices, each with a window. The two offices were almost universally laid out in a T-shaped arrangement, with a reception area large enough for a stenographer, files, and a waiting room, with the offices coming off this area.[45] Large businesses with multiple clerks and secretaries usually opted for more open floor plans. However, around 80 percent of tenants leased suites of less than a 1,000 square feet, or the equivalent of four or five office units.[46] This created a world of commercial real estate where, in the words of a late 1920s architect, "finance dictates the fenestration; rent rolls rule the parti."[47]

A skyscraper's enclosed space needed to be designed in a way that would allow people to use it, and the width of corridors and the placement of doors acknowledged that functionality. A corridor 6 or 7 feet wide separated the office units. When a corridor had offices on both sides, leading architects suggested that the doors not be opposite each other and that doors should be wide enough to allow a desk to pass through them, but no wider (this translated to a width of 40 inches). In addition, the number and placement of elevators should ensure that workers could easily access their places of work.[48]

Flow

The emphasis on movement within an office provided the conceptual space in which the vertical filing cabinet made sense. Advocates for the flow and circulation of work in offices sought to emulate the horizontal space of factories within the constraints that the verticality of skyscrapers placed on horizontality.[49] This perspective privileged a definition of system as a track rather than as an organizational focus on people.[50] By the 1920s this had become a conscious attempt to mimic the horizontal logic of the assembly line, with elevators assisting the vertical movement between floors that was rarely needed in a factory.[51] Advocates, usually enamored of the ideas of scientific management, believed an office should be organized to create a system for "forward movement," providing "sharply defined channels through which work can flow."[52] As L. C. Walker (of Shaw-Walker) put it, "Every time a piece of paper stops, a dollar is resting."[53] Productivity depended on the movement of paperwork to ensure the flow of information—not just the information required to complete orders but also information that could reveal the pattern of ordering, which would enable future production to be planned, not guessed at. Efficiency required information.

In attempting to mimic the assembly line's path from one "operation" to another, the office needed to manage the quantity of tasks, so that workers would not be waiting for others back down the "line" to complete their tasks. Advocates also considered the movement of people as they walked paper not only between departments but also around furniture and machines (notwithstanding the occasional implementation of pneumatic tubes and fantasies about the organization of desks to allow workers to pass paper from one seat to the next). On the rare occasion when these theories of movement were approximated in practice, it was usually in insurance companies.[54]

However, office floors that had standardized layouts and workflow systems were found more frequently in books than in buildings.[55] In the early decades of the twentieth century, even more rare was an entire building designed and built with ideas of efficiency and flow in mind. But the unique and the exception are so attractive that one example, Frank Lloyd Wright's Larkin Building, appears frequently in histories of the office and commercial architecture. Designed in 1903 for the Larkin Soap Company in Buffalo, New York, the building reflected Wright's intention to create a structure that was shaped by the work that occurred within it. This was work centered on the increased value granted to information; for Larkin, it signaled a move from a company that made things to one that marketed things. To free up the floors for paperwork, Wright set the staircase towers apart from the main building, a decision that also reduced the boxiness of the building. The result was a six-story structure with an atrium at its center, which allowed light to enter from above as well as from the sides. Wright's vision for the building depended on specifically designed furniture. Metal cabinets for the filing of index cards with customer information were built into the walls. Cushioned seats were cantilevered from desks on swivels so that they could be folded away for easy cleaning—an

THESE TWO GUIDES TO THE "WORK-FLOW" OFFICE ARE YOURS

★ The term "Work-Flow" applies to all Art Metal equipment. Briefly it means that every piece of Art Metal equipment is designed to increase the efficiency of the job for which it was intended—clerical, stenographic or executive. Work flows faster, more efficiently because there is a minimum of lost motion . . . hence the term "Work-Flow".

The booklets illustrated above tell how Art Metal desks and filing equipment operate to increase "Work-Flow" — how time and money can be saved by equipping offices exclusively with Art Metal.

These two manuals have been prepared for office managers and buyers because their chief job is discovering ways and means to save time and money.

With Art Metal's Airline desks in 22 styles and the complete line of Mainliner desks, almost every space and work problem can be solved. The Manual of Desk Drawer Layouts shows the proper desk arrangement for all principal executive and clerical positions, and the User's Guide to Office Modernization gives a complete picture of how to obtain maximum office efficiency.

Use these two booklets to gain increased "Work-Flow". They will be sent on request on your business letterhead.

★ CONSTANTLY AT WORK FOR IMPROVED OFFICE METHODS ★

Art Metal
1888 — 1940
STEEL OFFICE EQUIPMENT

Figure 1.4. During the first half of the twentieth century, the concept of work flow emerged to shape the design and organization of offices. This Art Metal advertisement from 1940 for office equipment catalogs introduces the improvements that consideration of work flow can bring to the office. Courtesy of the Fenton Historical Society, Jamestown, New York.

efficient design, but workers were not fans of the chairs' limited movement. Wright's architectural and design solutions intended to address the problem of the circulation of information in an office remained unique, but his awareness of the importance of furniture to work flow speaks to the most overt way in which concerns about the movement of information and people played out in offices.[56]

In contrast with this specifically designed office space by an iconoclastic architect, changes in the design and use of desks offer more pervasive and widespread examples of how the office was rethought as a site to facilitate the flow of paper. The emergence of a "flat-top" desk and the vertical filing cabinet offered a coordinated rethinking of the relationships between paper and work and between storage and retrieval.[57] The efficiency-based focus on flow and movement made storage a particular kind of problem. Ideas about work flow positioned storage as a sedentary state in which papers would take up long-term residence in a cabinet or desk. As

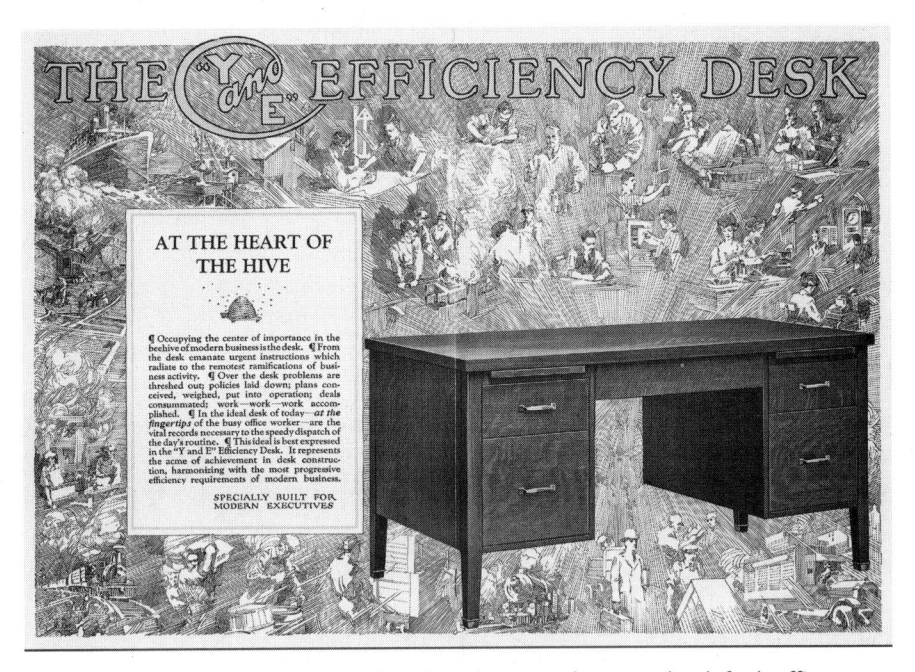

Figure 1.5. The flat-top desk, with reduced storage, became the default office desk in the early twentieth century. This two-page catalog spread promoting Yawman and Erbe's Efficiency Desk shows how it was sold as a way to make office work productive. Yawman and Erbe, *The Executive's Workshop* (1922), Trade Catalog Collection, Hagley Museum and Library, Wilmington, Delaware.

a leading proponent of office efficiency put it, "At best, any type of storage system is a passive agent in business."[58]

The increasingly complementary function of desks and filing cabinets emphasized the importance of the rethinking of offices as spaces organized to manage the movement of paper and facilitate access to information; this made storage primarily a temporal problem centered on retrieval. In this context, storage needed to take up as little space and as little time as possible. The solution was office equipment designed specifically to store paper on its long edge. This innovation was paired with a rethinking of the office desk as a "workbench."[59] In this new, early twentieth-century version, *"a desk should never be used for storage purposes* but should serve as a means for the prompt dispatch of business."[60]

The office desk had traditionally provided storage space for paper with drawers, pigeonholes, and spindles or spikes. This trend culminated in rolltop and high-top desks, which used pigeonholes and different-sized drawers to create elaborate partitions. These desks represented a specific approach to office work. They emerged in response to the problem of organizing space, not the problem of managing time, which took on increasing prominence only when the idea of efficiency began to reshape the business imagination.

Peak partitioning occurred with the Wooton Desk.[61] Patented in 1874, the standard model, with three hinged parts, had 110 compartments. Aside from numerous shelves, it included 40 pigeonholes, 10 drawers, and 11 different-sized racks for books. It was designed for an office where one person was responsible for multiple tasks that often required different forms of papers, writing equipment, and stationery. While they were moderately popular at best, the Wooton Desk Company's products illustrate the storage logic of a small office and the understanding of clerical work that, while subject to change, remained dominant through most of the nineteenth century.

Depending on their size, desks in the nineteenth century were designed for a one-man operation or for a clerk, the latter assumed to be a young man serving an informal business apprenticeship in a small office.[62] The clerk was imagined as a man who worked at his own pace, often hidden behind a desk stacked with pigeonholes, or, in the case of an accountant or bookkeeper, standing at a desk, where he would learn about the business in its entirety. The expansion of the geographical scale of businesses, and with it the increased use of paper, challenged the apprenticeship model of clerical work, notably through its division into specialized tasks (such as filing), at least in larger business concerns. This specialization was paired with a move toward greater centralization and control of clerical workers. The office emerged into its twentieth-century form in part as a result of the growing recognition that it was no longer possible for business knowledge to be kept in the mind or desk of one person, notwithstanding the efforts of Mr. Wooton.

The move from an "old-style storehouse model" to a modern flat-top desk foregrounded a change in the mode of storage—from personal to collective—as papers would ideally be moved from desks into filing cabinets.[63] In this shift the structure and design of a stand-alone filing cabinet represented a move away from the personal mode of storage associated with desks and any sense of possession office workers felt toward the paperwork they generated.

The aim was to prevent "valuable information" from being "scattered among the various members of the organization, who treat [it] as personal property and preserve [it] in their private desks as carefully as a squirrel hides his store of good nuts."[64] The long-term squirrel mode of storage would be separated from individual storage practices through the use of a filing cabinet and a desk that lacked storage space (and the centralization of information in the creation of filing departments and business libraries). From this perspective, "a desk is not a filing cabinet, and if anything is to be kept longer than a week, its place is in the files."[65] Acknowledging that it required "vigilance" to keep only the papers in use on a desk, experts celebrated the new flat-top desk as a way to assist the worker.[66] Instead of numerous drawers and pigeonholes creating a "receptacle for junk," a flat-top desk was designed to be a "repository for the storage of valuable, live material in use."[67] The only concern efficiency celebrants had about flat-top desks was that in some cases their tops could be so large that they forced workers to use excessive "reaching energy."[68]

The idea that paper was "live" spoke to the need to prioritize accessibility and retrieval to ensure that paper and information flowed through an office. This countered the idea that storage was passive or that paper once stored was dead, that it could not be found. Even when stored in a filing cabinet, it was placed standing to attention where it could be found easily in the future. In the words of one filing instruction manual: "In an office . . . the letters and other business papers ought to flow like a current past all the men who need to use them," but ultimately they had to flow "right on to the file clerk to be put away for future calls."[69] This raised a question: Where does one place a filing cabinet so that paper can easily flow into it through the hands of a file clerk?

Locating a Filing Cabinet

Thinking about where to put a filing cabinet prioritizes its dimensions, density, and exterior. It brings the structure of the cabinet to the foreground: What is behind it? What is underneath it? What surrounds it? How much space does a filing cabinet need? The standard four-drawer cabinet in the early twentieth century was about 51–52 inches high and 24–27 inches deep, although cabinet height was generally described in terms of drawers, not inches. Most early cabinets gained extra elevation from the addition of "sanitary legs"—sanitary because they allowed for cleaning underneath the cabinet and kept the contents of the lower drawers cleaner (as well as making them more accessible).[70] To be clear, a clerk's health was prioritized because it factored into her efficiency as a worker, not because of any concern for her general well-being. The relationship between efficiency and a clerk's body also determined the depth of the cabinet. In presenting "maximum capacity" as one of the seven essentials of office equipment, Shaw-Walker explained that a drawer greater than 27 inches in depth risked turning filing from an "arm operation" into a "walking operation," as the clerk would have to walk to the side of the drawer to access files at the rear. Effective use, in the form of easy access to all papers, determined the depth of a drawer and therefore the depth of the cabinet. Thus, in the design of filing cabinets, Shaw-Walker declared, "the vanishing point of utility is very close to 27 inches."[71]

A single four-drawer cabinet rarely stood alone. It was often placed next to other filing cabinets.[72] The main selling point of early filing cabinets was modularity. Absent that label, in the first decades of the twentieth century vertical filing cabinets went by a number of different names: unit, sectional, expandable, expansible. Office equipment companies intended the names to convey that offices could tailor filing cabinets to their specific needs, with

No. 1001
5-INCH LEG BASE
Sanitary Base for use under individual cabinets.

No. 1002
Same as No. 1001 but for use under end cabinets of a battery.

No. 1004
CASTER BASE
Increases height of a cabinet by 2½ inches.

Figure 1.6. These illustrations from a Steelcase filing cabinet catalog show buyers' options for adding bases to filing cabinets. The legs on these bases were often described as "sanitary" because they made it possible to clean underneath a cabinet. Courtesy of Steelcase Corporate Archives, Grand Rapids, Michigan.

the assumption that the businesses would grow, or at least produce more paper. Early advertising sought to communicate the ease with which filing cabinets could respond to the needs of a growing business. Recognizing this, most companies also made single-drawer units that could be the basis of expansion.

Side or top panels could be detached to facilitate expansion, with drawers added either side by side or stacked one on top of the other; the cabinets were designed to be connected with just a few bolts. Manufacturers claimed the resulting "battery" of cabinets would look like one seamless piece of furniture. Despite the emphasis on the vertical, most companies expanded filing cabinets horizontally. Efficiency drove the decision, but it was efficiency determined by labor, not space. The introduction of five-drawer filing cabinets brought this tension to the fore. These cabinets were advertised as cost-saving measures, with taglines like "Add 25% to your filing space rent free."[73] Manufacturers also touted the increased height as a solution to congestion in the office: "Office jammed . . . people falling over each other . . . filing cabinets flooding the place?"[74] However, critiques of the five-drawer cabinet suggested a problem with pursuing verticality as a principle of efficiency—taken to an extreme, it could produce inefficiency.

Regardless of the number of cabinets grouped together, office management literature emphasized that workers had to be able to access the drawers easily. A Shaw-Walker catalog expressed this concern using a barrage of numbers, a common approach for an industry invested in scientific management. The text offered the statement that an office paid $1,000 in labor costs to operate forty drawers of filing (the office paid $4 per square foot in rent). In terms of space, a five-drawer cabinet would save $72 annually, based on a four-drawer cabinet using 72 square feet for forty drawers and a five-drawer cabinet needing 18 square feet less to accommodate the same number of drawers. However, those savings assumed that file clerks maintained the same level of efficiency whether they were using four or five drawers. Shaw-Walker did not believe this was possible because the people employed as file clerks were women: "The average height of women is less than men and the top drawer of a 4 drawer case is about as high as an ordinary girl can work to advantage. Labor is the greatest and ever present fact of expense in any business and it is poor economy to reduce its efficiency because

Figure 1.7. This cover of a Remington Rand promotional pamphlet features the five-drawer filing cabinets most manufacturers had introduced by the 1920s. Sensitive to the price of office rent in large cities, companies marketed these cabinets as accommodating more storage in the same footprint as four-drawer cabinets. Author collection.

of a theoretical saving in office space by piling letter drawers one on top of another out of convenient range of the operator."[75]

Promotional materials for five-drawer filing cabinets (and for many four-drawer cabinets) accentuated the height of the women who operated them to deny the possibility that the height of cabinets could produce inefficiencies. In these advertisements female clerks were drawn (or photographed) standing tall with impeccable posture to accentuate not only the rectilinear design of the cabinets but also their height; if necessary, high heels elevated their bodies

Figure 1.8. This Remington Rand promotional pamphlet shows the priority office equipment companies gave to ease of access in the marketing of filing cabinets. To highlight how all drawers were accessible, advertising images often depicted tall women (usually in heels) easily retrieving papers from the top drawer of a filing cabinet. Author collection.

so they matched the additional height of the cabinet. From this height a tall, erect woman could easily "master" the filing cabinet as she gazed at the tabs inside a file drawer, while the catalog put her body on view for the assumed male reader.[76]

Platforms not attached to a woman's body could also be used to pair verticality and efficiency, and possibly provide more stability for a female file clerk than that offered by high heels. Office equipment companies produced stools and ladders designed to help workers access file drawers. A woman seated on a high stool could see into the cabinet and reach into or retrieve files from the top drawer. A low stool could be used to improve access to the bottom drawer, to lessen the exhaustion that bending over might cause (the cabinet's sanitary legs also assisted by raising the bottom of the drawer from the floor). However, some critics viewed the use of a stool as a failed solution, as it created new inefficiencies. The extra "operations" the clerk needed to perform—fetching the stool, climbing on it, stepping down from it, and returning it to where it belonged—added up to increased filing time.[77]

In businesses that chose to expand their filing units vertically, clerks often used ladders to access the uppermost cabinets. This tended to be the case in large enterprises such as insurance companies, which often extended cabinets to the ceiling, though these were usually for drawers that stored index cards or folded documents. Even though Leffingwell championed verticality, he was a ladder skeptic. He recounted an example of a "girl" working at a correspondence school who fell from a ladder while retrieving lesson plans from a large set of floor-to-ceiling shelves (he also did not like shelves). In this instance, according to Leffingwell, the school followed his suggestion and replaced the shelving system with fifty four-drawer filing cabinets, which required only one file clerk to operate because of the time saved by not climbing (and falling from) ladders.[78]

Rather than concerns about access, however, it was the availability of unused office space that most frequently determined the location of the filing cabinet. Consequently, a cabinet could be far from a person who regularly used it, as recounted in an efficiency parable about a frustrated clerk published in the *Idaho Statesman* in 1917. Consistent with the genre, the stand-in for the efficient clerk, "Helen Blank," got promoted as a result of her suggestions to make

Metropolitan Life Insurance Co.'s Home Office Bldg., N.Y.City. Glimpse of the Filing Section. The largest outfit of steel filing cases in the world.

Figure 1.9. Companies, particularly in the banking and insurance industries, occasionally had floor-to-ceiling stacks of filing cabinets. These were made accessible by ladders, as seen in this General Fireproofing catalog page and promotional postcard from Metropolitan Life. Author collection.

the office more efficient, which included moving her desk closer to the filing cabinet.[79] Miss Blank's story highlights that access became important when someone took the time to think about where to place a filing cabinet. This incorporation within economies of access gave the filing cabinet the only status it enjoyed in the layout of an office. If this happened, it produced varying degrees of awareness regarding not only the space required to fit a cabinet but also the space needed to pull drawers open to reveal papers. In large credit bureaus the accessibility of cabinets included clear pathways for women telephone operators who wore headsets with long cords to allow them to move among filing cabinets to find the credit records necessary to answer inquiries.[80]

The limited light in offices was another factor in the decision about where to put a filing cabinet. Champions of file clerks argued that filing should be prioritized in the allocation of space because it required good light for the constant reading necessary to ensure papers were filed correctly.[81] Other office commentators were less sympathetic. They considered filing to be a job that involved "continuous action," not continuous eyestrain. They contrasted filing with accounting, which they believed to be work that did require good light (in subsequent chapters, I discuss the gendered work hierarchy that divided mental and manual work such that few people questioned the need for accountants to have good light).

Filing cabinets tended to be placed against walls, regardless of the number of cabinets and the size of the room. However, when multiple filing cabinets were placed in one space they might form a hollow square or be positioned back-to-back.[82] In a specifically designated filing room these configurations were often combined; for example, cabinets might be placed around the wall with additional ones back-to-back in a hollow center.

At the level of office design, there was awareness that a filing cabinet was part of a larger assemblage. Experts suggested that regardless of where an office placed a cabinet, there should be at least three feet of aisle space to allow someone to walk by a clerk as she worked on a fully extended drawer. The space designated for multiple filing cabinets introduced the problem of clerks simultaneously accessing drawers in neighboring cabinets. A clerk who had to wait for another clerk to finish using a neighboring drawer represented a "sacrifice of the facility and celerity of action that is the main object of filing systems."[83] The proposed solution to this problem was to arrange cabinets so that they alternated back and front. This allowed every unit to be consulted at the same time.

The organization of office space affected the placement of filing cabinets in another way as well. Filing cabinets could be used as partitions to demarcate departments in open-plan offices. The scarcity of natural or artificial light made cabinets a desirable alternative to floor-to-ceiling partitions, which tended to shut out both light and air. Therefore, the exterior of the cabinet, acting as a partition, divided office work in the same manner that the interior of the cabinet divided paper.

A battery of cabinets could also provide a barrier between workers and customers, with the top used as a counter for waiting on customers or a table for clerks to work on. If an office needed to, it could buy purpose-built counter-filing cabinets that housed three drawers; at 42 inches high, these were about 10 inches shorter than the average four-drawer unit. They came with counters attached, and, as one company described its cabinet, "the battleship linoleum tops provide a warm, smooth working surface."[84] Ever wary of accusations that their products made inefficient use of office space, office equipment companies argued that the space saved by not needing tables more than compensated for the loss of filing space in each cabinet.[85]

The filing cabinet's use as a counter or divider underscores the primary function of an office: to organize objects, workers, and tasks. It also emphasizes how important the idea of flow was in the reimagination of the early twentieth-century office and the effect this had on the objects placed in office spaces. While few offices likely replicated those that appeared in office equipment catalogs, ideas of flow and verticality shaped the space in which a filing cabinet was placed and therefore the cabinet itself.

At the turn of the twentieth century, verticality became an important organizational logic for capitalism. Recruited from a longer history that linked height and organization in other contexts, it was an important technique for imagining and organizing the commercial and corporate office as a modern work space. Industrial development

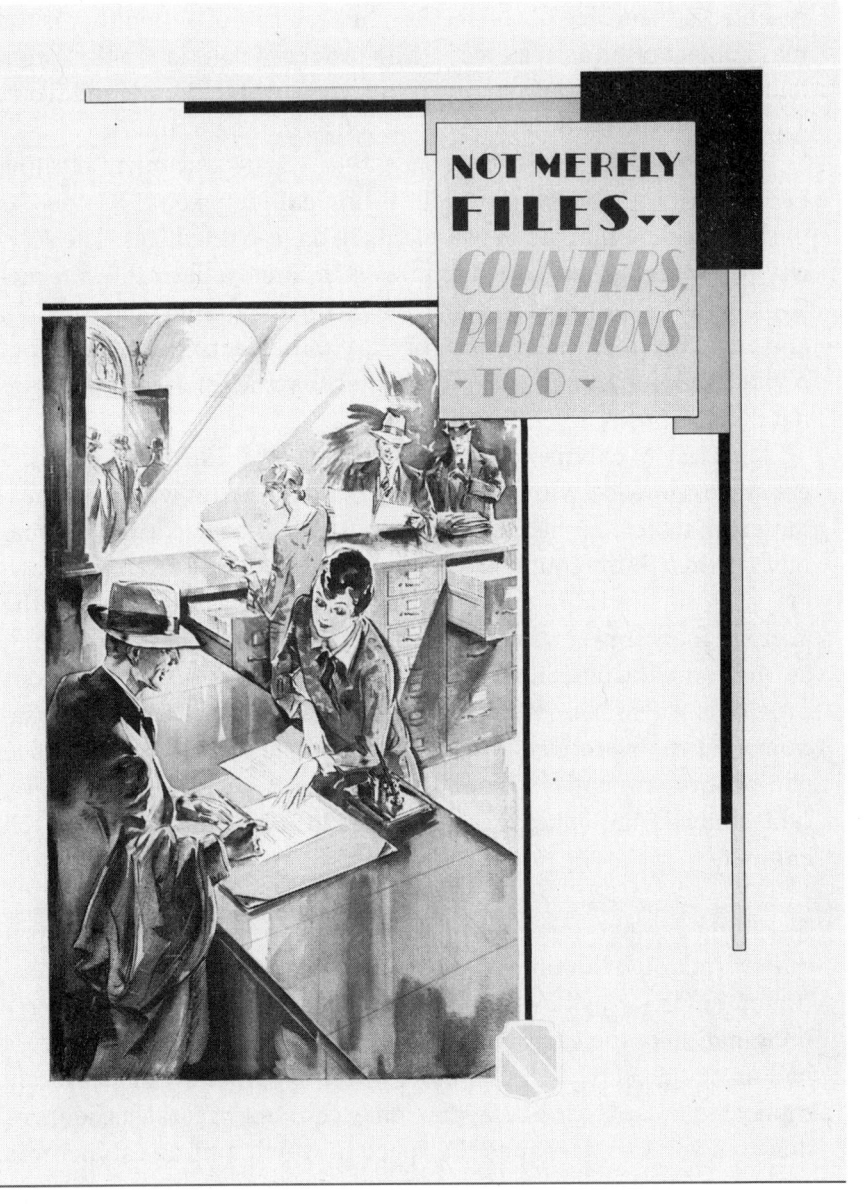

Figure 1.10. Shorter three-drawer cabinets were manufactured for businesses to serve double duty as counters for dealing with customers. As this Library Bureau promotional pamphlet shows, the clerk was almost always a woman and the customer a man. Courtesy of the Herkimer County Historical Society, New York.

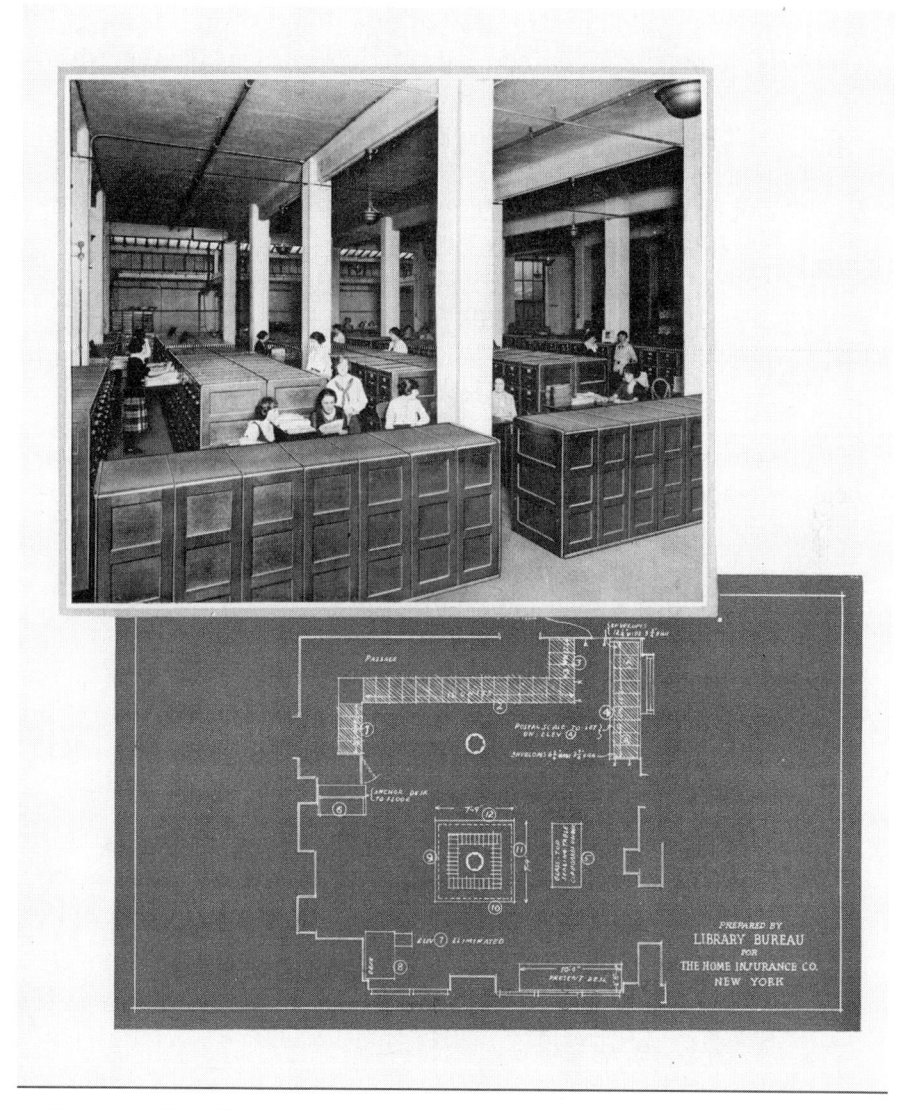

Figure 1.11. The Library Bureau used this image of the file department of a large Kentucky hardware company to promote its counter-height cabinets. The company used the cabinets to demarcate the space of the department. In an era when electric lights were not common or effective, the cabinets (unlike walls) prevented the obstruction of natural light. Courtesy of the Herkimer County Historical Society, New York.

triggered an obsession with growth that dangled the seemingly ever-present possibility of increased production. This fixation with growth made height a frontier for the colonizing logic of capitalism. The skyscraper and the filing cabinet suggest that there was an investment in using vertical structures as an architectural and design strategy to increase volume within a limited horizontal footprint. The filing cabinet was a vertical structure used to manage the horizontal space of the drawer, a rectilinear structure located in a vertical structure organized to manage horizontal floor space divided into offices.

Verticality was intended to create efficiencies of space and access. This investment in flow led to the emergence of cabinets designed specifically for the storage and retrieval of paper. Along with desks intended for specialized tasks and the development of standard paper sizes (e.g., for typing and for index cards), these cabinets acknowledged a move from "furniture" to "equipment." Equipment (along with "office appliances") represented a more functional and utilitarian approach to the office and office work, as well as the structure of equipment such as the filing cabinet.[86]

Chapter 2

INTEGRITY
Paper's Steel Enclosure

Storage is a label for a space of storing, where something is placed for future use. Storage technology claims and shapes space in particular ways, often through the creation of an enclosure, a place that secures the interior from the exterior. Storage depends on integrity to mark and secure the place where an object can be accessed. Concerns such as accumulation, continuity, and protection shape the creation of this space; thus, storage is not neutral.

For the filing cabinet, integrity guarantees ease of access and retrieval. In identifying integrity as a storage principle, I draw on its definition as the quality of being undivided and unbroken, of being whole. However, while the integrity of the filing cabinet explicitly includes its wholeness as an upright object, as a value in storage integrity can also capture anxieties about the moral uprightness of the relationship between what is stored and where and how it is stored.

Integrity speaks to concerns that greeted the filing cabinet's arrival in offices. As illustrated in Elihu Root's attempt to modernize the offices of the U.S. State Department, contemporaries viewed the invention of the filing cabinet as offering more efficient access to paper and information than had been available previously with the use of the book. Understood as a material container of specific information, the filing cabinet introduced a number of expectations related to the handling of loose paper. Access to specific information, the easy circulation of information on paper, and the ability to change and update the order of papers were some of the advantages that users projected onto loose paper. Using a filing cabinet could also produce concerns about misfiling that made visible the tensions surrounding storage, flexibility, and mobility, which these expectations created. Unbound papers stored in a folder in a drawer appeared to offer less integrity than pages bound in a book. It seemed a lot easier to misplace a sheet of paper than to lose a bound book, or to intentionally remove a sheet of paper from a filing cabinet without detection compared to removing a page from its binding.

To compare the integrity of a book with that of a filing cabinet is to recognize that the book and filing cabinet are both storage technologies for paper. As with Paul Otlet's contemporaneous use of index cards to enact the "monographic principle" in his bibliographic project of extracting content from books to separate content from authorial intent, placing loose paper in a filing cabinet was, in Ronald Day's words, "a curious extension of cultural assumptions about the book and a literal deconstruction of the book in its traditional physical form."[1] Day argues that such technologies provided a "more modern model for information technology that sees documentary forms as productive agents in the networked creation of information products and flows."[2]

The increasing faith in efficiency and its manifestation in the flow of paper brought worries about security and preservation that challenged the primacy that the use of a book had given to attachment. In addition to these more spatial aspects of integrity, the absence of binding raised more temporal concerns related to durability and permanence. Did the filing cabinet offer the same structural integrity as a book? Was the storage structure of the filing cabinet and the contents it held inside as durable as a book?

To examine integrity as a key principle in the storage of paper, I begin this chapter with the move away from the bound letter book and ledger book as important storage technologies in the office. I trace this transition through two overlapping developments: the introduction of loose-leaf ledgers and file boxes and the process of the construction of the vertical filing cabinet. The latter was increasingly presented as a guarantee of integrity, a means to assure permanence, access, and retrieval.

Loose-Leaf Ledgers

The emergence of the loose-leaf binder offers a perspective on integrity as a value attached to the storage of paper and on the stakes raised when the storage of loose paper challenged the existing conception of integrity. Concerns over the potential looseness a binder granted to paper instigated a series of legal challenges in the United States and Europe that focused on both the legal definition of a book and ontological conceptions of a book.[3]

Andrew Piper argues, "Books are things that hold things."[4] Specifically, a book is designed to hold sheets of paper as pages. In his analysis of Renaissance rhetoric, Walter Ong contends that the codex, with the arrival of print, saw a new investment in the spatial imagery of a book as a container, as a volume; it provided a way to contain the profusion of paper.[5] While it could be used to hold other things (bits of paper, flowers), a book was not structured for "loose and insertion-friendly arrangements."[6] The value of the book as a storage technology became its structural cohesion, its wholeness. In the bound book the page has merit because it is attached on the inside, it is not paper on the outside where it might be lost. That is, the bound book has integrity because of collation, because of a perception of finality.

The definitiveness associated with the book shaped its function in offices, where it was used to store relevant information that owners or workers copied from loose paper. Outgoing correspondence was often stored in copybooks, and financial information was written into ledgers and account books. Loose paper in the form of notes, checks, bills of exchange, bills of lading, and other warehouse receipts constituted the "vitals of business," as an early historian of loose-leaf storage labeled them.[7] However, while the content was critical for a business, the loose paper was ephemeral; the "vitals" became permanent when they were copied into a bound book to become a record.

The introduction of the loose-leaf ledger did not stop the practice of copying information from smaller pieces of paper, but it did provide a mode of storage not based in bound paper.[8] As with all aspects of modern filing, the celebration of the merits of the loose-leaf ledger centered on how it could group all records about one thing in a single place. This proved appealing as it allowed a business to record and organize accounts by customer name as opposed to recording transactions in the order in which they occurred; pages could be inserted when needed. In a loose-leaf ledger the alphabet trumped the calendar.[9]

In the era of efficiency the binder repurposed the book by introducing the "mechanical" into the world of paper. In making this argument Cornelia Vismann describes the binder: "Unlike books, the inside contains, in addition to mere letters and numbers, a gripping mechanism attached to the cover that spears, staples, and if required, releases papers."[10] This file mechanism "bound" the papers

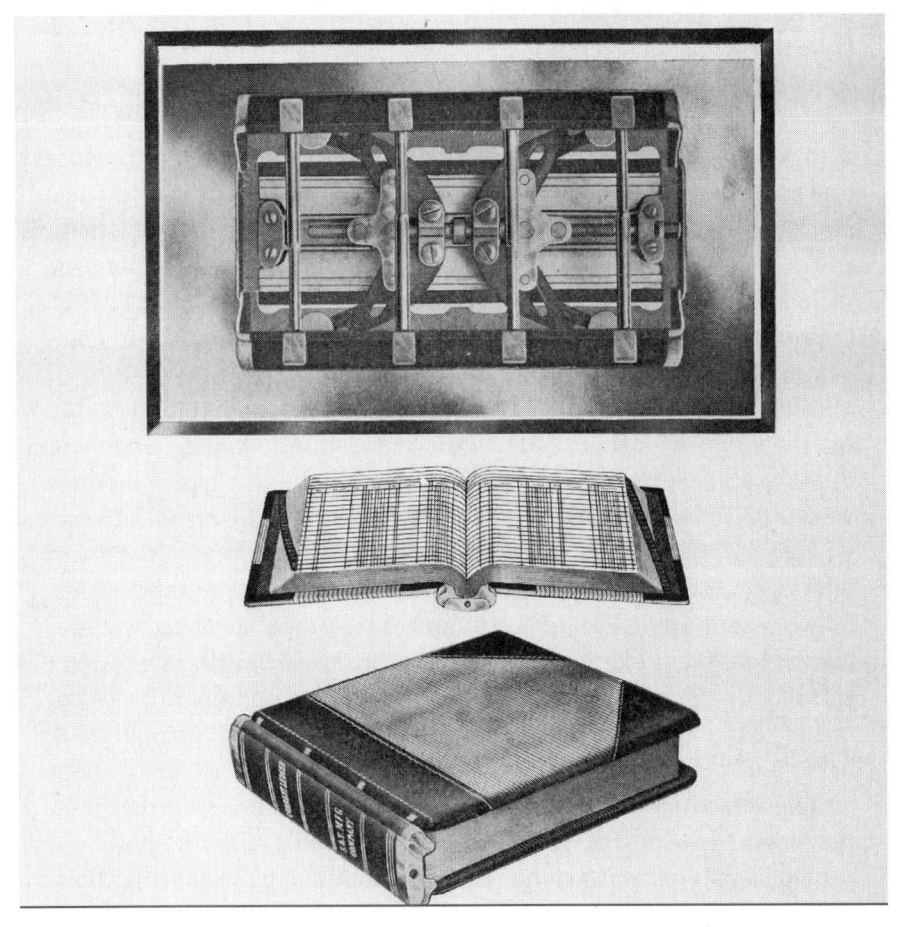

Figure 2.1. This page from a 1907 book on how to use loose-leaf ledgers celebrates the mechanisms that store ledger paper in a binder. Despite such treatises, loose-leaf binders as ledgers raised concerns about the loss and manipulation of loose papers. Courtesy of the Hathi Trust Digital Collection.

in an attempt to give integrity to the binder and, therefore, the papers within it. The binder offered a mode of storage that enabled the specificity and rearrangement of the particular that efficiency demanded; that is, it maintained the discrete material presence of individual leaves of paper in an effort to satisfy the new expectations in industry, commerce, and law for the timely retrieval of particular bits of information. The bound book could not satisfy these expectations, for, as Markus Krajewski argues, "a book cannot ever provide loose and insertion-friendly arrangements in alphabetical order; glue holds together those things that, according to the dictates of time, belong together."[11]

Although the loose-leaf binder seemed to speak to increasingly dominant ideas of speed and precision, when it began to be used as a ledger in the 1890s it encountered the ideas of a preexisting dominant social institution.[12] Specifically, courts had to decide whether a loose-leaf ledger could be accepted as evidence. The repurposing of the book as a binder introduced uncertainty over whether a loose-leaf ledger was in fact a *"book of accounts"* as law required. It looked like a book—it had a cover and a spine, and it contained paper. However, the monopoly of the bound book as an accounting technology had produced the assumption that what made a book a book was its binding. That is, the structural integrity of a bound book had become accepted as proof that the accounts contained within it were original records and, therefore, had been created "at or about the time of the transaction."[13]

A book, through the permanent binding of pages, was understood to provide integrity in the sense of *an undivided or unbroken state*. Binding created a sequence of pages. Chronological order was applied to this seriality to provide the space for the daily recording of accounts. That is, organization by calendar supported the temporal claims that underwrote originality and authenticity: that accounts were recorded in a timely fashion. Further, courts believed that a person disciplined by the obligation that bound pages (enclosed in a book) created to record events regularly would not "deliberately . . . contrive to mediate a fraud against his neighbor."[14] Legal precedent granted power to the presence and structure of a book through the assumption that using a bound volume to record transactions produced a commitment to update records on a regular basis. The authority placed in collation, in the finality of a bound book, made the book's contents impossible to change or revoke; a bound book was irrevocable.

Thus, the introduction of loose-leaf ledgers created an opportunity for a rigorous debate about attachment and looseness in paper storage (and therefore a challenge to how integrity would function as a storage principle). Presented as evidence in a courtroom, the loose-leaf binder seemed to weaken the evidential authority of accounts, if integrity was defined by material wholeness. Compared to the assumptions that had coalesced around the book, loose leaves, not permanently secured in a book, appeared to invite the irregular recording of transactions. Unbound papers, untethered to a system, created the possibility that a person might, "in the heat of passion," create, alter, or destroy a record.[15]

The appearance of the loose-leaf ledger illustrated the extent to which legal reasoning had linked material wholeness and moral soundness, such that the former was interpreted as evidence of the latter. In the creation of records, bound paper offered evidence of moral integrity; legal reasoning interpreted it as offering freedom from moral corruption. If a binding could not prevent a page being removed, it would likely provide evidence of that removal or absence through residual paper in the gutter or a gap in page numbers. The probable appearance of this evidence was viewed as a further deterrent. In contrast, loose paper too easily invited the presumption of loose behavior. That the leaves were loose, and thus easily attached and detached, raised doubt and suspicion about what could happen to individual sheets of paper.

These initial concerns were dealt with in a series of legal decisions over the first half of the twentieth century, through which the loose-leaf ledger became recognized as a book of accounts. Specifically, a binder came to be accepted as satisfying the requirement that in the presentation of accounts "the whole must be taken together" in documentary evidence.[16] To be evidence of the whole of an account, the relationship between integrity and completeness had to be adjusted to accommodate a mode of storage in which "each page is separate and distinct" such that an account could be assembled to exist separately from other accounts.[17]

A "book" came to be defined as a technology of enclosure over and above any claim to permanent attachment. According to courts, the loose-leaf ledger achieved the status of a book because "sheets of accounts are bound together in a folder or assembled in a cover with such a degree of permanence that a book results."[18] To be clear, no ruling ever clarified the precise degree of permanence required for a binder or folder to constitute a book. However, with a deft touch, the permanence previously associated with binding could now come from evidence of a *continuity of action*. Courts implicitly granted loose-leaf ledgers the required permanence if they were shown to be part of recognized record-keeping systems. System or routine took over the disciplinary responsibility previously given to bound pages. As long as the system of recording accounts remained consistent over time, a loose-leaf ledger was understood to provide the permanence associated with a bound book of accounts.

A 1946 ruling clarified that the definition of a "book of accounts" depended on this specific understanding of enclosure and permanence but also made clear that the reinterpretation of integrity depended on intention and motivation. This decision, upheld by a state appeals court, rejected a plaintiff's statement that twelve sheets of paper fastened by a staple were from his "loan account ledger" and, therefore, records created as part of his regular course of business over a period of nine years. The court also used the absence of preprinted ledger sheets as evidence that the business did not have a regular system of accounts. In its ruling the court stated, "Notations made upon loose sheets of paper are not accorded the presumption of accuracy and reliability which they have when entered in book form and are therefore inadmissible as books of accounts." It then clarified, "Sheets do not become an account book by being fastened together at the top by means of a staple."[19] A "book" was a recognized place of storage, an enclosure that formed part of a continuous and regular record-keeping system. In providing an enclosure, it marked a place designated solely for business records. Setting aside a place to store records provided evidence of intent for a system of record keeping. Therefore, legal reasoning had reached a consensus that system and technology could discipline paper and the people who handled it.

These decisions put the evidential authority of a "book of accounts" in line with the provisional permanence that modern paper storage technologies provided to offices in the name of efficiency. The flexibility, the ability to rearrange contents, meant that while the place of a piece of paper might change in the binder (or the filing cabinet), the classification system fixed its relative location. This offered a version of integrity that was mutable. It gave a collection of papers the quality of wholeness while acknowledging that completeness was never permanent. A completeness that remained provisional spoke to the potential growth and expansion expected in a business that followed the principles of efficiency.

Therefore, the book's integrity, its value as a technology that bound paper, had become a problem in an office that approached time as something to be saved. The particularity of a book was insufficiently granular, notwithstanding that within a book pages are individuations, especially in contrast to scrolls.[20] This particularity had been further enhanced with the addition of indexes, chapters, and page numbers, which set up the book for "discontinuous reading" or "consultation."[21] However, as Secretary of State Root's championing of vertical files over bound volumes shows, while the contents of the book can be consulted as something discrete to counter the continuity offered by its bound pages, there is no such discontinuity in storage; a request for one letter brings you the entire book in which it is stored. Root's faith in vertical filing illustrates that the desire for the particular is even more evident in the emergence of the filing cabinet than in the use of the loose-leaf ledger. In the absence of any form of binding, loose papers in a manila folder could be easily accessed and isolated according to specific requests; their material form aligned more with the particular information a person sought. However, like the articulation of the loose-leaf ledger and a record-keeping system, the filing cabinet was understood to secure a method of classification that attached loose paper to a system; this also brought concerns about attachment and integrity, albeit ones that did not make it to the courts.

Cabinets in Nineteenth-Century Offices

Cabinets designed to store and provide access to unbound paper appeared in offices in the decades immediately prior to the invention of the vertical filing cabinet.[22] These structures were to be used with one of the new filing technologies introduced to manage the growing volume of loose paper, which had increased for a number of reasons as office practices changed, management emerged as a profession, and offices adopted the typewriter and carbon paper. In the last quarter of the nineteenth century, filing developed with the invention of the box file, flat file, and document file. The box file was a cardboard box shaped like a book (although popular, these were rarely stored in cabinets). As the name suggests, the flat file (also known as the board file) stored paper flat on a board. The more ambiguously named document file stored folded papers upright on their long edge.

Figure 2.2. In the middle of the nineteenth century, two common forms of paper storage involved pigeonholes and spikes. These illustrations from an office equipment catalog contributed to the company's argument that older storage technologies failed to protect papers and made it difficult for workers to access individual papers. Author collection.

Until the middle of the nineteenth century, paper in offices tended to be found in bound volumes (placed on desks, on shelves, or in cupboards), in desk drawers, in pigeonholes, and in piles (either loose or fastened with a buckled strap or string) on top of desks or shelves. Papers were also impaled on the sharp-pointed wires of spindles or spikes, which stood on desks. With the exception of the book, these were all open modes of storage. They exposed paper to vermin and dust. They often failed to respect paper, instead creating opportunities for damage, such as when clerks stuffed papers into pigeonholes or skewered them on spikes. Flat files, box files, and document files addressed these concerns in different ways. The cabinets designed to store large numbers of these files also functioned as initial indexes; drawer fronts had labels that provided grids intended to make it easier for workers to locate groups of papers.

A box file explicitly emulated a book. A heavy cardboard box, with a hinged cover and sized to fit "commercial paper," it was designed to be opened like a book. Pages or pieces of manila paper, tabbed with letters of the alphabet, were attached to the box along their edge; papers were filed between these sheets. Manufacturers

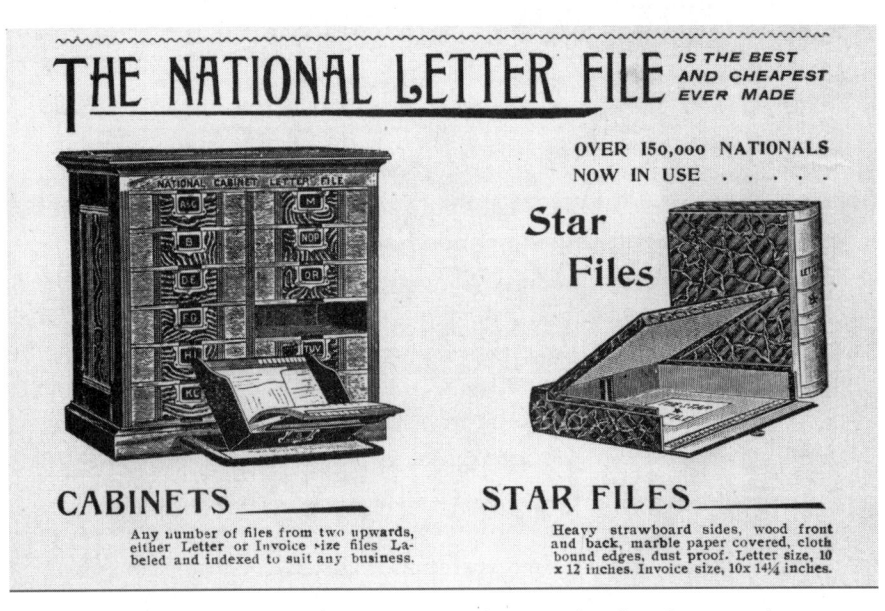

Figure 2.3. This nineteenth-century National Letter File advertisement shows the box file. Modeled on a book, it could be turned into a drawer and placed in a cabinet. Author collection.

produced box files with leather-covered backs and other surfaces coated with marbled paper to accentuate the comparison to books. Once a box was filled, it was labeled with a beginning and end date. A closed box was usually stored like a book, vertically, on shelves or desks, but as discussed shortly, it could be stored in a cabinet.

Patented in 1868, the earliest flat file (or board file) was a tray with one built-up end; it was sometimes described in office literature as a single open "drawer."[23] It had a compressor spring to keep papers in place and to allow them to be "released with a single movement of the hand."[24] Clerks usually stored flat files on preexisting shelves or on desks.[25] However, in the middle of the 1870s flat files began to be "arranged drawer fashion in a wooden shell."[26] The three-inch-high tray front took on the function of a drawer front, with a label added to indicate the contents of the file "drawer." The book continued to provide a model for use. The compressor spring attached to the inside of the flat file's tray functioned somewhat like the spine of a book, allowing the attached papers to be turned like bound pages. In extolling the usefulness of a flat file, an inventor noted, "The free edges of the papers above the one that is to be examined may be thrown to one side in the same manner that a book is opened."[27]

The most popular version of a flat file was the Shannon File, which was unique in its use of an arch to connect papers to the board. In a structure apparently "familiar to the entire commercial world," when the arch was opened letters could be filed on two upright posts.[28] Although this required punching holes in papers prior to filing, and thus demanded additional labor, proponents claimed the closed arch provided "absolute security against accidental loss."[29] Along with this "secure binding," another selling point was the ease with which the Shannon File could be carried around the office, even when an individual file held four hundred letters.[30] Large files usually came with tabbed index guides to assist in the location of particular documents.

When not being carried around the office, the Shannon File was often stored on a desk or hung on a nail attached to the side of a desk. Proximity to a desk underscored the belief that this file offered a useful way to store frequently consulted papers. Instead of being lost after being left on a desk in piles, papers such as unfilled orders could be found easily in the storage space the Shannon File

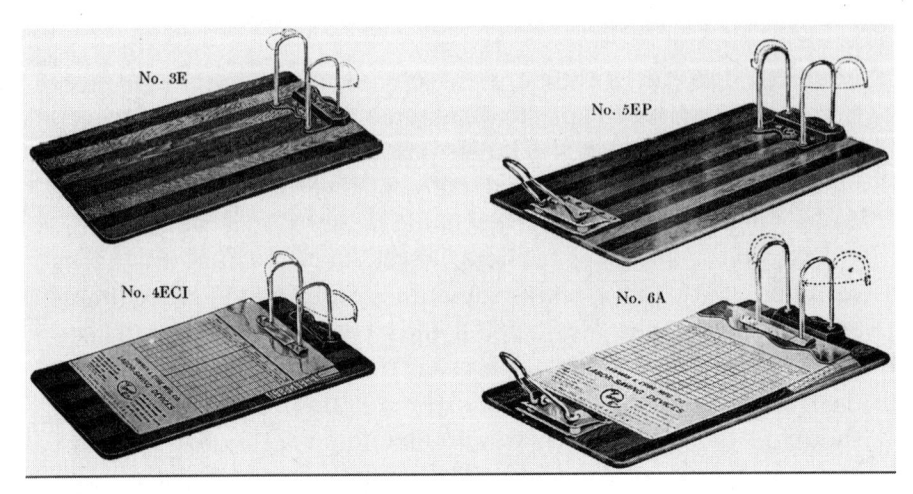

Figure 2.4. Flat or board files were commonly used in the late nineteenth century for storing small quantities of loose paper. This page from a Yawman and Erbe catalog shows the most popular board file, the Shannon File, with its signature arch that allowed more papers to be stored. Yawman and Erbe, *Filing System Supplies* (1923), Trade Catalog Collection, Hagley Museum and Library, Wilmington, Delaware.

Figure 2.5. This page from an 1887 Schlicht and Field catalog illustrates how a Shannon File could be turned into a drawer to be stored in a cabinet. These early files were indexed to provide easier access to specific papers. Courtesy of the Smithsonian Libraries, Washington, D.C.

created. Offices with a large number of Shannon Files could store them in specially designed cabinets that could hold up to ninety-six files. To make them work in a cabinet, Shannon Files that were stored this way had facings attached to their ends to provide the "drawer fronts" that other board files had.

Document files began to appear in the early 1880s, particularly in legal offices. They are the ancestors of today's magazine file: a flat box cut in half along the diagonal. The box was stored vertically and had a spine like that of a book. However, because their design allowed loose papers to be stored and individually retrieved, these files avoided the limitations of storing paper in bound books. Made of wood, a document file was designed to store larger documents that had been folded into a smaller, more standard size; in filing, the term *document* would come to refer to paper folded for the purposes of storage.[31] Once folded, the papers, with annotations describing their contents added to the outside, were placed upright in the box. A clamp-like device attached to the back end of the box supported the papers. By releasing the clamp, a clerk could make a file's "contents fall back by their own gravity, aided by the retractile force of the paper."[32] As with flat files, manufacturers began to offer cabinet structures for offices that needed to store large numbers of document files; these were particularly popular in legal offices, where from the 1890s the use of document files had become increasingly common.

Cabinet System

Manufacturers presented cabinets as a way to manage large numbers of individual files. They sought to associate cabinets with the growing interest in "system." However, "filing by system" struggled to gain acceptance. Potential customers viewed cabinets as a "commercial luxury" until well into the 1890s.[33] Initial cabinet designs were very basic. The early flat filing cabinet had a top, a bottom, and three enclosed sides to provide storage. The fourth side revealed shelves divided by vertical partitions that created "recesses," "cells," or "compartments" for flat trays or drawers and eventually document files. Whether it was called a "shell," a "frame," or a "cabinet," the use of that three-dimensional spatial division paired with drawers, trays, or boxes marked these cabinets as an important

No. 13 Globe-Wernicke Units.

No. 13 D. F. Unit, showing File with Suspending Device.

A UNIT fitted with our regular Globe Document Files of proper size to accommodate papers $4\frac{3}{4} \times 11\frac{1}{4}$ inches in files with Suspending Device, and $4\frac{3}{4} \times 11\frac{3}{8}$ inches in files without this device. Unless ordered, files will be sent without Suspending Device.

Net delivered price, without Suspending Device, $10 00
" " " with " " $12 00

Subject to terms on page 4

The Globe Document File.

Patented Feb. 14, 1888.

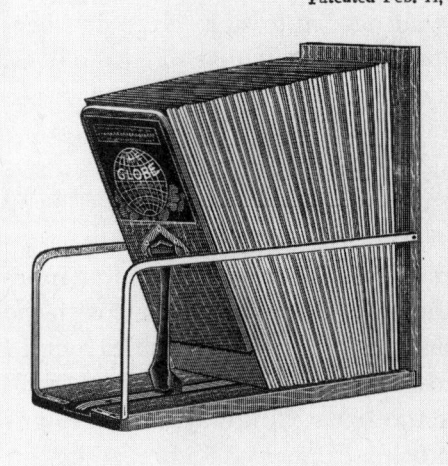

FOR folded papers the "Globe Document File" in quality, appearance and adaptability to the purpose for which intended, is without a peer. Its action is easy and simple, the clamping device or compressor direct and positive, no springs or complicated mechanism to get out of order and so thoroughly constructed that it cannot become inoperative through wear. The fronts are made of Quartered Oak, the bottoms and compressor boards of hard wood, the pulls and label-holders of polished brass, and the interior fittings of metal, nickel plated. These files are largely used in attorney's offices, court houses and public buildings and are fast coming into popular favor for general business purposes.

Figure 2.6. This page from an 1899 catalog for Globe Co. filing products shows an example of a document file. In this form of storage, papers were folded and placed in a drawer with rods and a clamp to keep them standing. Globe Co., *The Globe-Wernicke Elastic Cabinet: A System of Units* (1899), Trade Catalog Collection, Hagley Museum and Library, Wilmington, Delaware.

development in paper storage for offices—something different from books and spindles but also a change from pigeonholes.

By the 1890s the typical cabinet consisted of a base (sometimes in the form of a horizontal drawer) with pilasters flanking the body of the cabinet and a top or cornice.[34] Catalogs offered cabinets in multiple sizes and styles, accompanied by descriptions that emphasized the consistency and unity of the structures—their structural integrity. Table or shelf cabinets stored anywhere from two to a dozen "files." Freestanding cabinets, for document or flat files, were usually capped at 75 files. The Globe Company's 75-drawer flat file was 76¾ inches high, 69½ inches wide, and 15 inches deep, containing drawers that were 5 inches wide and 15 inches deep; the entire cabinet weighed 730 pounds, without papers.[35] Making use of specially designed bases, these cabinets could be combined to hold as many as 480 "file drawers." In response to concerns about appearance and privacy, most manufacturers offered optional doors or slatted fronts that could be rolled down.

The cabinet structure highlighted the ongoing importance of integrity to paper storage: it provided a space that, through the wholeness of its structure, constituted and maintained a system. In a "cabinet of files," the grid layout of partitions and drawers presented the user with a collection of files visibly subordinated to a larger system (this cabinet logic is the focus of chapter 3). In one of its catalogs, the Globe Co. highlighted this layout to explain the utility of its cabinets: "It is in fact a single huge file, systematically arranged, embracing hundreds or thousands of subdivisions."[36] The cabinet system produced order by subdividing the interior space within its frame. The subdivisions created specific locations for individual drawers or boxes. These smaller enclosures became the base unit for the storage of loose paper (a role the folder took on in the vertical filing cabinet). Surpassing what could be achieved with pigeonholes, these divisions were then further divided by sheets of manila paper attached to the interior of the wooden file. In another catalog the Globe Co. clarified the value of the subdivisions within a cabinet system: "The chief advantage in referring to papers by the cabinet system over the single file system is in the more perfect subdivisions of matter; so that in a properly arranged cabinet, any desired paper among 40,000 or more, can be more readily found and referred to than the same paper among 500 in a single file."[37]

While labels on file drawers or file boxes allowed the fronts of the entire combination of files to be seen at a glance, accessing the files often created problems that threatened the integrity of the cabinet system. The use of wooden files as drawers introduced a new set of problems for the storage, retrieval, and protection of loose paper in

Figure 2.7. Large wooden cabinets were produced for different kinds of flat files. As this page from an 1898 Globe Co. catalog illustrates, they often had roll fronts for added privacy and security. Courtesy of the Smithsonian Libraries, Washington, D.C.

an office. As an inventor noted in a patent application that offered a solution for such problems, drawers could stick or, more commonly, fall out, causing "an accident that always results in inconvenience, and frequently injury to the contents of the drawers."[38] Horizontal drawers that formed the bases of cabinets caused similar problems. Therefore, the last quarter of the nineteenth century saw numerous patents issued for stops to prevent drawers from falling out of cabinets and friction-reducing devices to assist with the smooth movement of drawers. Because these problems were not unique to office cabinets, the solutions could be applicable to (or borrowed from) the drawers of bureaus, card catalogs, clothes presses, washstands, and any other cabinet that required sliding drawers.[39]

Manufacturers' desire for "convenience" also saw them literally

Figure 2.8. Large cabinets or wall units presented a grid that showcased the system used to organize papers in offices that emphasized the specific place files had and their relationship to other files. This illustration from Schlicht and Field's 1887 *Labor Saving Devices* catalog shows a cabinet for five hundred document files. Courtesy of the Smithsonian Libraries, Washington, D.C.

balance access and integrity as they sought to ensure that drawers could be accessed while remaining part of a system, rather than being completely removed from the cabinet and becoming individual stand-alone files. Supporting devices allowed drawers to be tilted, inclined, or suspended while remaining attached to the cabinet.[40] An 1884 catalog described the ideal operation of a cabinet in its promotion of the company's products:

> We have arranged these FILES into NEAT, CONVENIENT and HANDSOME Cabinets of a variety of styles and sizes and connected them with the Cabinet in such a manner that the File can be withdrawn from its place and still remain attached to the Cabinet . . . and the papers [can] be examined without removing them from their place in the File or detaching the File from the Cabinet; but if it's desired to remove the File to a table it can be readily lifted from its place.[41]

The decision to highlight the ease with which paper could be removed without detaching the file from its place in the system spoke to the anxiety over attachment and system that characterized the emergence of loose-leaf ledgers. The importance granted to the correct place for papers illustrated how integrity became tied to a structure that gave unbound paper a proper place, a structure that provided propriety for the storage of papers.

Not all manufacturers agreed with this approach to propriety and integrity. One office equipment company argued that "suspension files" were useful only for drawers at eye height. The alternative the company offered, borrowed from library card catalog cabinets, was a sliding shelf placed at a convenient height between two rows of drawers so a file could be completely removed but remain adjacent to the cabinet and its location in the filing system. However, the company did acknowledge that suspension files could be useful in public offices and insurance offices where workers used ladders to access files stacked to the ceiling.[42]

Linked to the idea of system, the cabinet came to be viewed as more useful and practical than a shelf as a place to store groups of papers. Searching through bundles stacked one on top of the other on a shelf was now regarded as time-consuming. However, shelf advocates liked to point out that retrieval time was actually increased

Figure 2.9. This page from a 1910 Yawman and Erbe catalog offers another example of the grid structure of larger cabinets. This type of cabinet accommodated document files and the different flat drawer files in use immediately prior to the uptake of the vertical filing cabinet. Yawman and Erbe, *Record Filing Cabinets* (1910), Trade Catalog Collection, Hagley Museum and Library, Wilmington, Delaware.

if a worker had to open doors, unlock a cabinet, pull out a drawer, slide a partition, or raise or let down a flap. In a somewhat new development, which was harder for shelf lovers to counter, dust had become a problem in the office, and dealing with it required enclosed storage spaces with easy-to-clean exteriors. The emergence of dust as something to be dealt with represented another way the office had become modern. According to an early twentieth-century British clerk, dust and other "matter out of place" on the walls and ceilings of an office had once suggested "solidity, respectability, and age." However, where dust had previously been seen as evidence of a firm's long-term existence, such that clients "breathed an air of financial stability," in the modern office, the concern was that through dust and dirt people breathed in something much less desirable; this was especially problematic for those people who were expected to work efficiently in an office.[43] Therefore, like anything out of place, dust did not belong in a modern office.

The Integrity of a Filing Cabinet

In the 1890s, as cabinets for flat files and document files found their place in offices, the vertical filing cabinet began its hesitant emergence before quickly becoming the paper storage container of choice in the second decade of the twentieth century. The vertical filing cabinet's exterior and interior were different from those of previous cabinets, which produced new concerns about access and retrieval, and about the cabinet's integrity as a space of storage for paper. Manufacturers actively sought to negate such anxieties by championing the physical integrity of the vertical cabinet (the provisional permanence that file drawers provided through folders and tabs is the subject of the next chapter). As Shaw-Walker's "Built Like a Skyscraper" campaign illustrates, a common strategy was to link the construction of a filing cabinet to ideas associated with progress and the modern; by 1910, steel had become the material of choice for filing cabinets.

At the turn of the century, steel was "the modern construction material" through its association with railroads, skyscrapers, and automobiles.[44] This "age of steel" made it "the logical material for office equipment."[45] In the early 1920s, an office equipment catalog for the Jamestown, New York–based Art Metal Construction Company

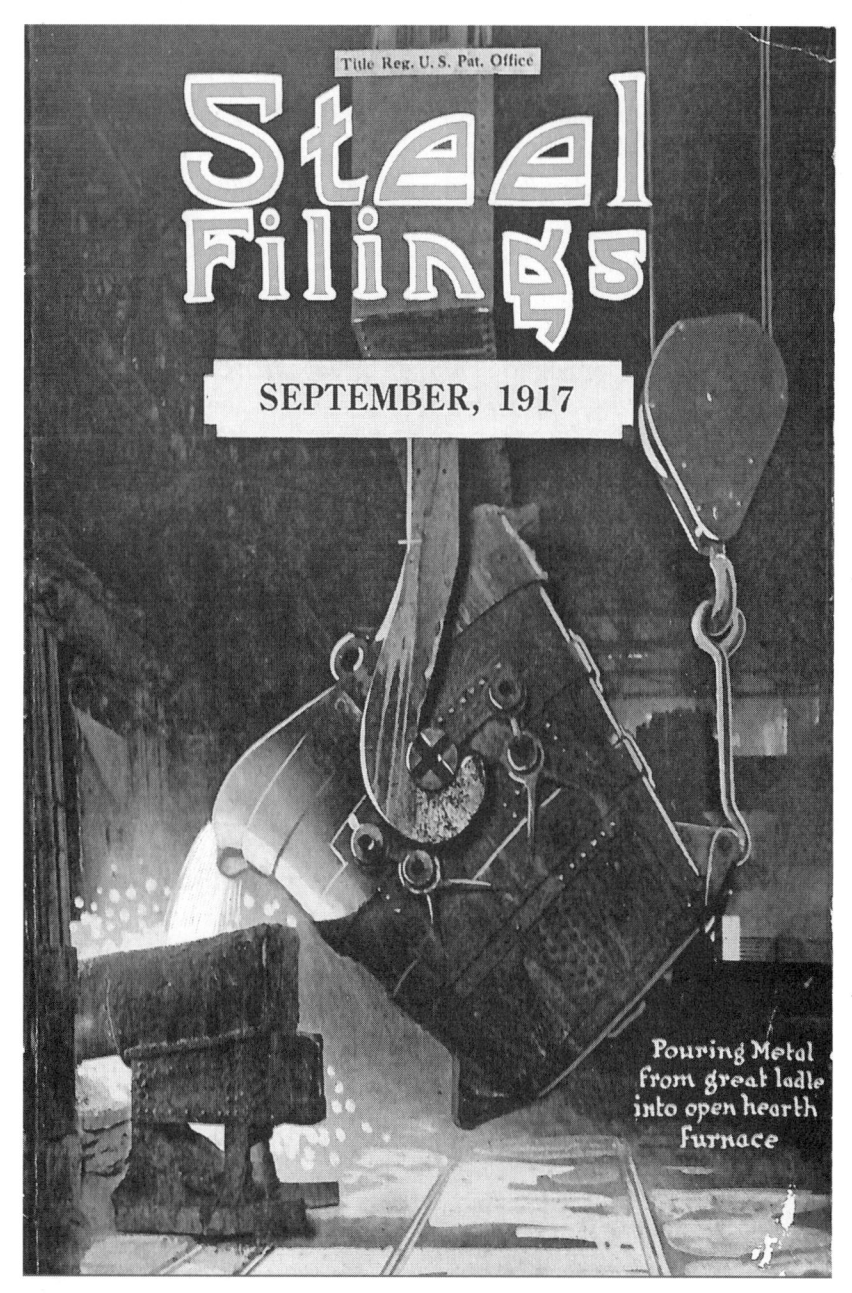

Figure 2.10. This cover from a 1917 issue of Art Metal's company magazine highlights electric furnaces in the manufacture of steel. When office equipment companies began to manufacture filing cabinets from steel in the first decade of the twentieth century, it was a relatively new material, still subject to debates about the best way to manufacture it. Courtesy of the Fenton Historical Society, Jamestown, New York.

proudly located the manufacturer's founding in the era of steel. In this history at the beginning of the twentieth century, "steel was replacing wood wherever possible" (ships, railcars, buildings), as "wood had been tried and found wanting. Steel had proved efficient in every way . . . the efficient business man of the future would be demanding efficiency in everything."[46]

Established industrial prophets and inventors also considered steel a necessary material for the furniture of the not-so-far-away future. In 1911, Thomas Edison told *Cosmopolitan* that steel "is destined soon to fall from its high pinnacle as the skeleton of skyscrapers, to become the material of which furniture is made." Reinforced concrete would replace steel in buildings and, because it was lighter, steel would replace wood in furniture. Edison also claimed the amount of steel used in furniture would cost about one-fifth the price of wood. In addition, speaking like a metal office equipment salesman, Edison noted that "polished steel takes a beautiful finish," offering a "perfect imitation" of stained wood. He finished with the confidence of an appointed prophet as he declared, "Babies of the next generation will sit on steel high-chairs and eat from steel tables. They will not know what wooden furniture is."[47] Edison noted that one unnamed New York firm was already making steel office furniture (he was likely referring to Art Metal).

In fact, by 1911 several companies were making steel office furniture and equipment; some had been doing it for at least a decade. Companies like Baker-Vawter and the Library Bureau had begun as wooden office furniture manufacturers; others, such as Ohio-based Van Dorn and General Fireproofing, were metal manufacturers who diversified into office equipment. However, while the move away from wood was retroactively presented as inevitable, steel furniture met initial resistance. Salesmen found early steel cabinets laughed off as "tin files," with one file clerk purportedly saying in disgust, "They have given me these cans for my department."[48] To many customers, steel office furniture and equipment "greatly resemble[d] in appearance a bulky office safe or an unshapely kitchen-range."[49] In the first decade of the twentieth century, this would have been an accurate observation more often than not, as steel office equipment frequently had problems such as unattractive finishes and drawers that would not open easily.[50]

Office equipment companies were not initially familiar enough with steel to signal a guaranteed future for the young industry. Prior to this period, the use of metal could impede any design, especially compared to wood, which was a material that manufacturers were comfortable working with. It took at least a decade for them to recognize that they needed specific kinds of steel for different parts of a cabinet, to determine effective ways to shape and fabricate the steel, and to recognize the need to weld rather than use rivets and bolts.[51]

The office equipment industry's initial failure to use steel effectively was not surprising; steel had to exist as such before it could be used in the manufacture of a filing cabinet. Steel in its modern form was as new as the office equipment industry itself, and its ascendancy had been far from certain. As Thomas Misa argues, the eventual domination of steel was the result of a convergence of factors, including scientific research, industrialization, innovation, the tensions between private industry and government policy, the rationalization of factories that produced steel, and the demands of the railroad, building, and automobile industries.[52]

In the 1880s and 1890s, Frederick Taylor initially attempted to apply his ideas of efficiency, what he would later label scientific management, at Midvale Steel and Bethlehem Steel. The latter emphasized high-speed steel, the need to cut steel faster, which was a response to the growing demand for steel from U.S. railroad companies during the second half of the nineteenth century. The decision to prioritize maximum production rather than maximum quality drove the adoption of the English Bessemer method of steel production. This method produced steel by blowing steam or air through molten iron; the process depended on extreme heat to melt or fuse the product.

The importance of the Bessemer method to U.S. railroads and the country's developing steel industry led to a scientific debate over what exactly steel was, a debate that had significant consequences for pricing and import classification. A fusion classification definition replaced the initial definition of steel that used carbon content to differentiate steel (0.2 percent to 1.0 percent) from wrought iron (0 percent) and cast iron (2 percent). By 1900, metallurgical textbooks came to accept the definition that had been honored in daily practice for a couple of decades: if metal had been completely melted, it was steel (with a strong chemical component used to classify varieties of steel).

However, the question of how to make steel a pourable fluid remained unsettled. By the early twentieth century, the demands of different industries for different types of steel marginalized the Bessemer method as preferences developed for an open-hearth technique and then the use of electric furnaces. Urban development and the use of steel in buildings fed the demand for open-hearth steel, which was considered more effective for structures than Bessemer steel. Electric furnaces provided overwhelming competition for open-hearth steel, largely driven by the automobile industry's need for alloy steels (steels that contained small amounts of elements other than carbon and manganese).

As the production and use of steel changed, the fledgling office equipment industry was able to offer products that more closely fulfilled the claims made in its promotional literature. The successful construction of steel filing cabinets and the comparison to skyscraper construction hinged on welding, which facilitated a manufacturing process that, for all intents and purposes, made each cabinet a singular piece. The industry quickly coalesced around a process that built a cabinet using a skeleton frame of pressed steel that was welded together, with sheets of steel welded to the top, bottom, and sides of the frame; hence all steel filing cabinets were "built like a skyscraper." Although not all companies exploited the link to the skyscraper, the filing cabinet industry did make the manufacturing process central to its claims regarding the permanence, protection, and rigidity of its products.

The marketing of steel filing cabinets stressed how the metal was worked so it could support the structural integrity of a filing cabinet. This emphasis possibly came from the assumption that people were familiar with wood and the basics of woodworking, but not with metalwork.[53] However, the importance placed on the production process also contributed to the representation of filing cabinets as modern. Not only were they "built like a skyscraper," they were also efficiently constructed through standardized steps. As one company stressed in a puff piece published in *Banker's Magazine,* in its new factory "everything [was] installed with the single view to make the best steel filing section with the least 'lost motion' and the greatest possible chance of a hundred per cent average."[54]

Catalogs for steel filing cabinets paid detailed attention to the production process. This usually took the form of a two-page photographic spread at the beginning of the catalog. Each photograph illustrated one step in the process. Collectively, the images portrayed the construction of a filing cabinet as an assembly line of standardized and consistent steps. To describe each step, catalog writers made a point of using technical terms that were usually, but not always, defined. Steel was "pickled"—placed in tubs containing sulfuric acid—to remove scale. The steel sheets and parts were "annealed" and "cold-rolled" to remove possible strains and stiffness; these steps involved expert knowledge about the right balance between temperature and time. Sheets were then exposed to a machine to create a flat surface called "stretcher-level" or "patent-level." Welding was described in detail, be it "electric welding" for flat joints or "oxy-acetylene welding" for corner joints. The finishing process, by which the cabinets took on their olive-green or wood-grain appearance, received considerable attention (suggesting lingering concerns about the appearance of steel cabinets). This multistep procedure involved primers, baking enamel, copal varnish, oils, and sanding. The application of these coats was interspersed with at least three eight-hour sessions in a 250-degree oven. According to one company, olive green became the default finish for metal filing cabinets because "a warm olive green, neither drab not dull . . . is a neutral tone which harmonizes with any surroundings."[55]

A cutaway image of a cabinet's interior usually followed the catalog description of the manufacturing process. Parts were carefully labeled, with the drawer slide receiving particular attention. Similar cutaway images often appeared in catalogs for wooden filing cabinets, but these were accompanied by noticeably fewer details, with little information about woodworking techniques or the production of the cabinets. One small photograph might be included, but if space was devoted to construction, it was usually to explain the kinds of wood used for the cabinets and drawers.[56]

The large, oiled sheets of first grade furniture steel, made to exact specifications, are cut accurately and carefully to size in large power shears.

The hand-hammerer with his expert touch finishes up a perfect sheet, after the cold-rolling has removed bends and buckles.

In these huge presses, the sheets—oiled to prevent any rusting—are for[med] into shape between their powerful ja[ws] as many as five men working on one p[ress]

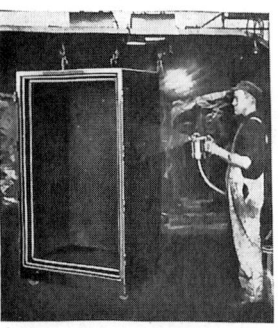

A spray coat is put on the more bulky products which would not permit of a dip coat, after being thoroughly cleaned and dried, of course.

This is a section of a press-room where the smaller parts are formed and punched out. It shows particularly the forming of the slides for the filing cabinet drawer suspension.

Making the
the Protecti[on]

The experience o[f] facts and figures o[f] and statistics of laws, creeds, and [] agreements and con[di]tions and individua[l] ently h[as]

After this second baking, making the cement and second coat as part of the safe, it is sanded. This operation further smooths down the surface, after which, another paint coat and baking.

Here we have the assembling of a desk. All the parts entering into it have been accurately made and have been painted except for the final finish.

The assembling of a four-drawer [vertical] letter file. It is held in a vise-like [grip] keeping it absolutely in shape wh[ile] suspension, drawers, etc., are fit[ted]

Figure 2.11. These pages from an Art Metal catalog illustrate the different stages involved in the manufacture of steel office equipment. This common representational technique sought to associate office equipment with modern assembly-line production practices. Courtesy of the Fenton Historical Society, Jamestown, New York.

vets or bolts—after forming, the product is made like one piece by welding. Here is a drawer being electrically welded, insuring maximum strength.

With the larger pieces, such as the famous Underwriter's Label, fire-resisting lightweight safe, it is welded by the oxy-acetylene blowtorch.

The oil and dirt are next thoroughly taken off by benzine, and the pieces wiped. With the smaller pieces a dip coat is now given for the ground color.

Equipment for Your Records

as gone before; the nd failure; the data g compilation; the es of mankind; the en nations, corpora- preserved and effici- rt Metal.

The first coating of paint is baked on in these large ovens. Each "run" is timed, and the temperature carefully regulated to insure an even, thorough job.

The baked-on first coat is gone over, and every small crack or dent in the surface filled with a special preparation of cement. This it is given another coat of paint and baked.

comes the final finish, and here a craftsman is putting on a circassian graining—rivalling Nature's own. ishing coat of varnish, a last thorough baking and—

It stands ready for inspection and final adjustment—all parts must work smoothly and accurately, finishes must be up to the standard sample, and the work must pass rigid inspection.

Crated up in sturdy wooden frames, and wrapped with heavy paper to prevent scratching and marring, it goes to the shipping room and is checked directly on the cars that stand at the doors.

Protecting Papers

Direct comparisons to wood were critical to manufacturers' initial attempts to sell steel office equipment. Steel filing cabinets arrived with the declaration that wooden cabinets were defenseless against the extremes of weather: damp or wet weather would cause drawers to stick; hot weather would cause cabinets to warp. As one manufacturer put it, "Steel is indifferent to fierce temperature and rack and

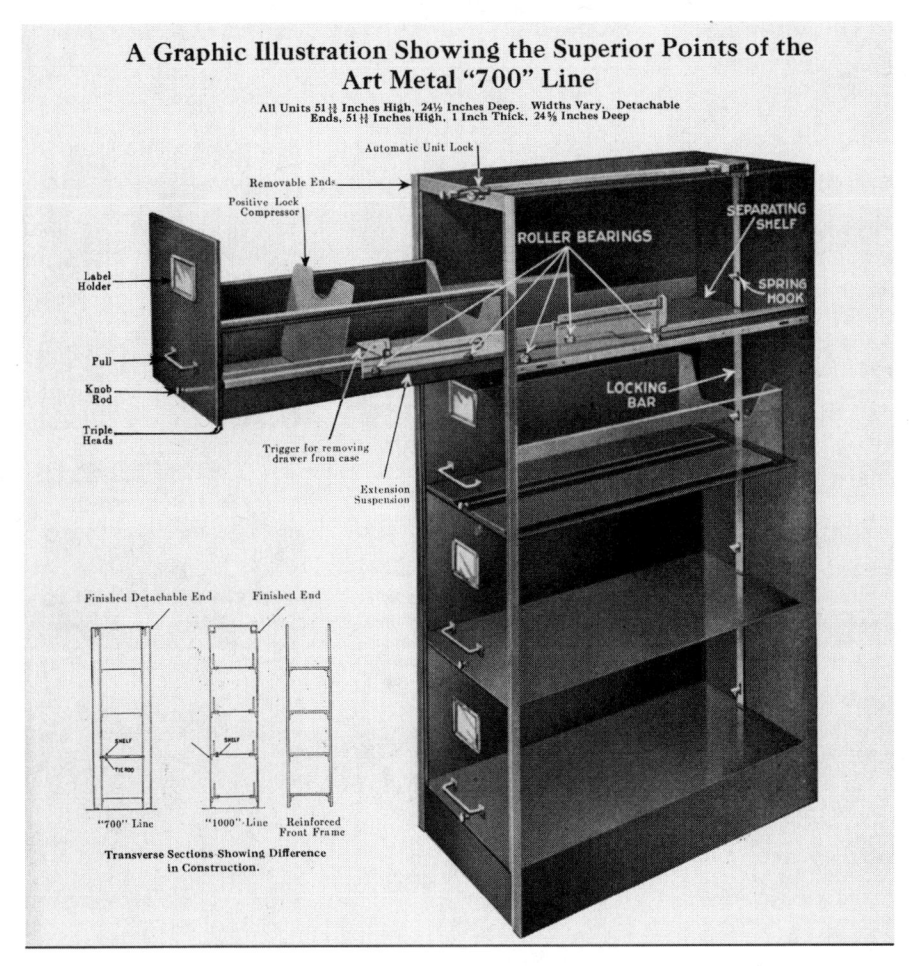

Figure 2.12. This cutaway image of the interior of a filing cabinet from an Art Metal catalog was intended to illustrate that cabinets were built to ensure the smooth operation of drawers and therefore easy access to files. Art Metal, *The Book of Better Business* (1917), Trade Catalog Collection, Hagley Museum and Library, Wilmington, Delaware.

wear of time."[57] To shore up wooden cabinets against claims that steel cabinets lasted longer, manufacturers attempted to reduce the costs involved in making wooden cabinets.[58] A no-frills four-drawer steel cabinet with an olive finish sold for forty to fifty dollars; different finishes or drawer slides and better-quality steel could increase the price to more than seventy dollars. Wooden cabinets were usually ten to twenty dollars more expensive than steel cabinets. One way in which manufacturers made wooden cabinets cheaper was by removing the horizontal frames between drawers. However, this created the impression that wooden cabinets were not sturdy, which reinforced the idea that steel filing cabinets provided more integrity than wooden ones.[59]

The efforts to sell steel cabinets focused on heightening their distinction from wooden cabinets by defining storage primarily as a problem of protection. The protective characteristics of permanence and indifference to surroundings were key advantages of steel over wood. Office equipment companies did not shy away from discussing the need for protection. Two years after the end of World War I, one company presented the office as a battlefield:

> Always in times of warfare have the guards been *guarded*. The army with its outposts, the outpost with its sentinels, the sentinel with his gun. Every day shows the need of adequate protection for those things that should guard business from danger. The man who best protects himself from danger and loss is the best business man.
>
> Your business is or *ought to be guarded by its records and papers*. They can only protect you against loss and danger *as you protect them*. It is logical therefore, that those things that guard you most should be best protected....
>
> GF Allsteel Files . . . will protect your records against fire and weather, against dirt and vermin. They mean efficient protection.[60]

As this advertising copy suggests, promoting steel filing cabinets involved highlighting a number of threats. Steel could provide a successful defense against "rats, mice, and other vermin." In the world of metal office equipment these creatures found easy access to wooden cabinets, "littering the receptacles and imperiling papers stored for protection."[61] However, according to many advertisements, fire presented the main danger to paper in offices. Attempts to protect files from fire aligned the emergence of the steel filing cabinet with the ongoing development of office safes.

Manufacturers quickly encountered the problem that steel filing cabinets did not offer their paper contents real protection from fire. Since steel does not burn like wood, an initial argument that steel provided better fire protection than wood seemed like a no-brainer, at least for salesmen. But while steel does not burn, it does conduct heat. In a fire, a steel filing cabinet's interior could get so hot that the paper inside would burn or at least be singed so badly that its contents would become illegible.[62] By the 1920s, the industry-wide solution for fire protection was to add a second wall of steel to a cabinet, with a double layer of asbestos between the steel sheets. While this might have protected papers, the introduction of asbestos occurred at the same time research in the United Kingdom began to suggest the dangers of this material for workers. Corrugated sheets of asbestos came to replace roll-board asbestos, because corrugated sheets interrupted airflow as "air space [was] broken into small cells."[63] This sensitivity to airflow and the need for protection caused "monolithic construction" to become important in the marketing of filing cabinets.[64] The idea of a unit as a singular piece supported claims not only for strength and rigidity (as in the skyscraper comparison) but also for preservation, as (through welding) the cabinet provided a "a monolithic insulation body, with no joints to open up for heat to do its insidious work."[65]

Not all companies were sold on simple asbestos. Art Metal used what it called "terrasbestos," which the company described as a "naturally formed compound of diatomaceous earth and asbestos fire . . . [that] possess[es] the strongest and surest fire resistance known to Nature or Science." Befitting the genre of promotional literature, an unnamed "eminent expert," a scientist, was quoted in support of this claim: "The walls of Hell must be of terrasbestos to keep the fire in."[66] The Safe-Cabinet Company ignored asbestos completely but adhered to the borderline humbuggery of promotional prose. According to its catalogs, the insulation of Safe-Cabinet products derived from "the scientific regulation and release of chemical reagents which counteract the heat, yet are not released until fire comes. They have no deteriorating effect upon Drawer-Safe contents."[67]

STEEL FILING EQUIPMENT

Fire-Wall Steel Filing Cabinets

Steel–*plus*–Asbestos is Better than Steel Alone

Unequaled Rigidity

Our method of framing, bracing and electrically welding the steel channel members of the framework insures the highest structural strength and rigidity. Even the trial by fire in the testing furnaces failed to distort any part of the frame.

Smooth Welded Corners

This is a feature your file clerk will appreciate. Our acetylene welded smooth corners make scratched hands and torn clothing an impossibility. This construction makes the front and folded edge of the drawer one solid piece. It's a "finished job."

Detachable Ends

Only one pair of steel-plus-asbestos ends is needed for any battery of Fire-Wall Filing Cabinets. They are easily placed in position or removed for the alignment of additional cabinets by simply adjusting four thumb screws on the interiors of the end cabinets.

Single Motion Compressor

One motion with one hand pulls up or releases the sturdy free-acting compressor inside the drawer. Our equalizer bar on the compressor brakes prevents any possibility of jamming or binding under use or abuse.

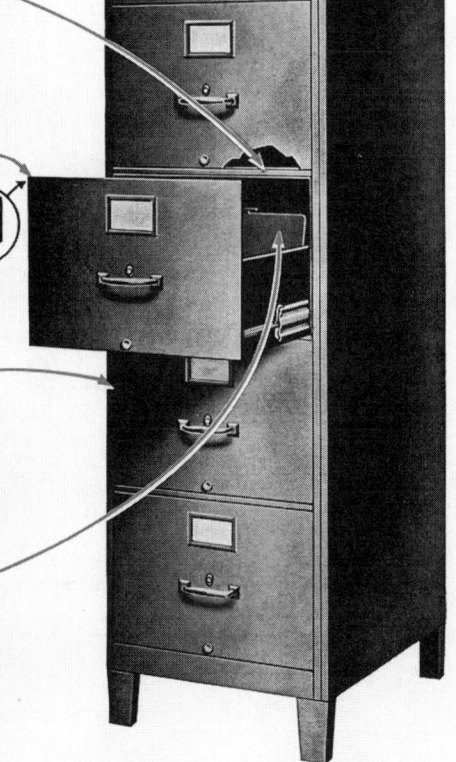

Figure 2.13. This page from a 1922 Yawman and Erbe catalog shows the important role asbestos played in the office equipment industry's attempts to protect papers from fire. Although companies claimed this solution was based on comprehensive research, the dangers asbestos posed to workers had yet to be considered in any significant way. Yawman and Erbe, *Steel Filing Equipment* (1922), Trade Catalog Collection, Hagley Museum and Library, Wilmington, Delaware.

Safe-Cabinet claimed to be the first company to develop steel storage equipment that provided legitimate fire protection, a feat it accomplished by studying the "habits of fire." Founded in 1905 as a small manufacturing shop before becoming a corporation the following year, Safe-Cabinet produced publications recounting a history that included its being the first company to send a product development manager to work for the New York Fire Department. By the 1910s, Safe-Cabinet had established the "Furnace Test." In such a test, as the company described it, a safe would be placed on piers in a large, room-sized furnace so that flames could reach all sides of the safe: "Electrical thermometers register the furnace temperature as it steadily climbs. In 10 minutes 1400 degrees Fahrenheit is reached—the melting point of glass—in one hour 1800 degrees Fahr-

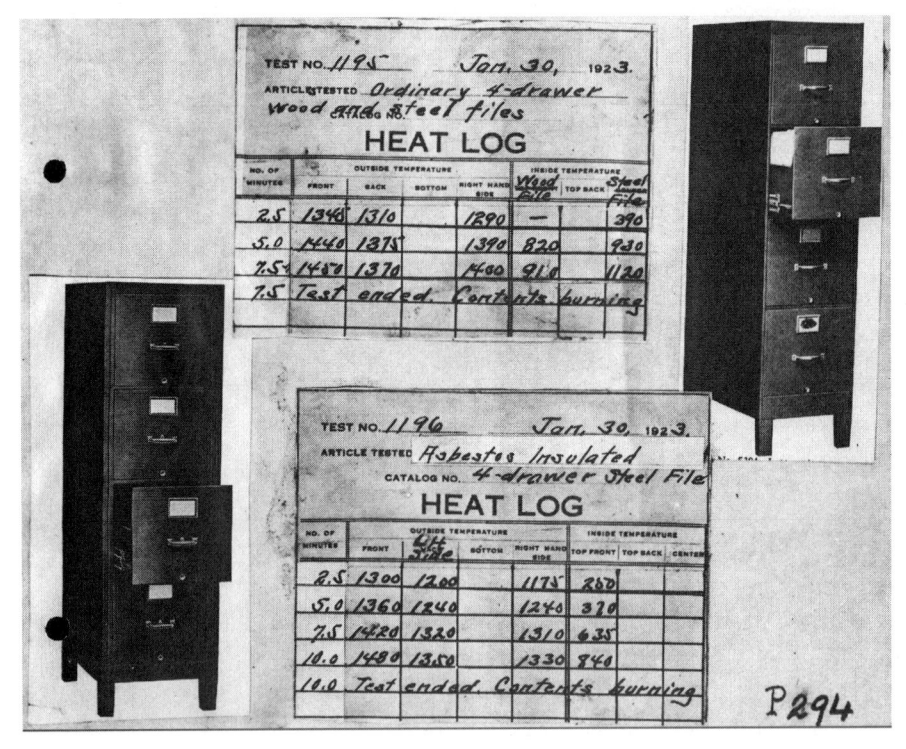

Figure 2.14. Office equipment companies promoted the research behind the production of fire-resistant filing cabinets. This Safe-Cabinet catalog page shows the "heat logs" that recorded how different cabinets fared when exposed to heat in large room-sized furnaces. Safe-Cabinet Co., *Wood and Steel Files* (1924), Trade Catalog Collection, Hagley Museum and Library, Wilmington, Delaware.

enheit and brass has melted, in two hours 1950 degrees Fahrenheit, and three hours 2,000 degrees Fahrenheit."[68] These claims appeared in a brochure in which the text was inserted in the middle of full-page color drawings. The surrounding illustrations presented the test as occurring on a theater stage with an audience, some members of which were brought to their feet, perhaps by the excitement of the test, or possibly to get a better view. The spectacle of fire was also deployed at the annual sales conventions of major safe and filing cabinet manufacturers, where "dramatic tests" provided high points.[69]

While these spectacles played their role, the industry came to embrace the work of Underwriters Laboratories, described by the *New York Times* as "a topsy-turvy land wilder than anything Alice in Wonderland ever dreamed of—a place where experts burn and smash and break the creations of other experts, all to make life and property safer."[70] Founded with funding from the National Board of Fire Underwriters, the laboratories ran tests that provided insurance companies with information so they could base their policies on something other than guesswork. A product that went through the testing process was given one of three possible grades: A, B, or C, with the lowest grade still considered satisfactory for inactive records. The tests conducted on safes and cabinets followed the furnace tests introduced by Safe-Cabinet. In one type of test, a red-hot filing cabinet with papers inside was dropped thirty feet onto an uneven pile of bricks and then sprayed with water from fire hoses. This was intended to replicate the reality that many fires resulted in the collapse of buildings. Office equipment promotional literature frequently invoked the use of Underwriters Laboratories tests, with some companies even specifying the UL grades on their labels to support their claims about the "strength" and "rigidity" of their products—aspects of what I am calling "integrity."

Although office equipment companies used fire to play up fears of loss, they were less concerned about theft. Therefore, locks received limited attention in catalogs and other filing literature. When they were mentioned, they were usually promoted as offering privacy rather than security. Most companies offered two types of locks for filing cabinets: flat key locks for individual drawers and single "automatic locks" that secured all the drawers at once. The latter were paracentric locks, with five tumblers that all had to be operated to open the locks. In catalogs, the short explanations devoted to locks usually noted that this feature made the automatic locks harder to tamper with or pick. However, Art Metal's catalogs promoted these locks as reinforcing the extent to which its cabinets protected papers from fire. Being able to lock all of a filing cabinet's drawers easily was important because a fully locked cabinet was more likely to maintain its structural integrity in a fire, especially if the building collapsed and the cabinet dropped several floors to the ground. Whichever option the customer chose, locks represented an additional cost. Most steel filing cabinets were stamped to allow for the installation of flat key locks for individual drawers, at a cost of one or two dollars per drawer, but knocking out a hole for the installation of an automatic lock could cost twelve dollars, adding almost 20 percent to the cost of the cabinet.[71] As their name suggests, automatic locks were promoted as easy to use; a cabinet could be locked at the push of button and opened with only a quarter turn of a key. Locks spoke to concerns about retrieval as much as to the need for protection.

Access: Drawer Slides

The integrity of a filing cabinet promised not only protection and preservation but also access; integrity in part concerns structural attempts to guarantee retrieval and access. As with other new storage technologies for loose paper, such as loose-leaf binders, manufacturers and users valued integrity only to the extent that it enabled paper to be stored and retrieved quickly with minimal effort. The importance that manufacturers attached to ease of access and retrieval underscores that expectations for efficiency made a particular conception of integrity critical to the filing cabinet.

As the exercise routine in Shaw-Walker's "Built Like a Skyscraper" campaign illustrated, easy access to a filing cabinet depended on functional drawers (Figures I.5 and I.6). The Shaw-Walker campaign used the celebration of male athleticism to showcase how well the company's file drawers worked, with images of a man jumping into a drawer showing a cabinet's "strength" and a man hanging from an open drawer depicting "rigidity." The advertising copy made this explicit: "Do a hand stand on a loaded Shaw-Walker drawer. Even that will not impair its smooth, silent, speedy action.

The slide will still operate as freely as before. It will coast in and out at a touch."[72] These claims also featured in advertisements that targeted prospective salesmen, who were told to perform these stunts in person: "When the gymnastics are all over and your prospect thinks for sure the file is in for a month's repairs, you show him how swiftly, smoothly, silently the drawers coast open and shut."[73]

Other companies were more direct about the relationship between the cabinet and its drawers, reducing the cabinet to a structure that existed only to "guarantee perfect operation of the drawers."[74] That is, a drawer that could be easily opened and closed relied on the structural integrity of the cabinet to support its movement and keep it attached to the cabinet. Wood again became a foil in the celebration of the efficiency of steel. Promotional literature claimed that the latter produced drawers that readily opened because, unlike wooden drawers, they "cannot swell, shrink, warp, and never stick."[75]

Ease of access depended on smoothly operating drawer slides, "the

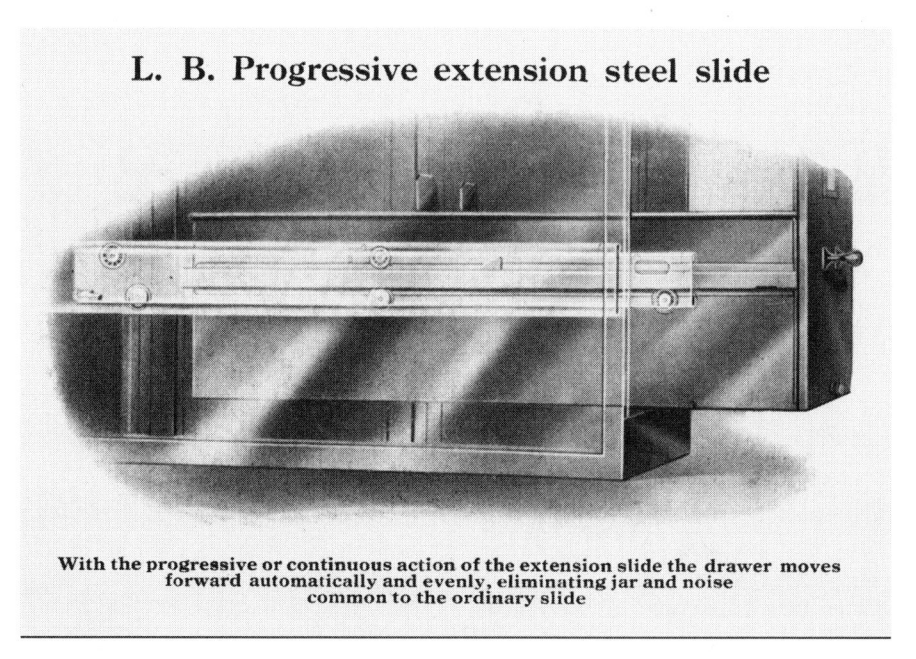

L. B. Progressive extension steel slide

With the progressive or continuous action of the extension slide the drawer moves forward automatically and evenly, eliminating jar and noise common to the ordinary slide

Figure 2.15. The drawer slide was critical to the promotion of a functional, efficient filing cabinet. This illustration from an early 1920s Library Bureau catalog promoted the continuous action of a drawer slide by naming it "progressive." Author collection.

life of a steel filing cabinet."[76] Early drawer slides were the sources of a number of problems related to the movement of drawers in and out of steel cabinets. Drawers would fall out of cabinets when pulled out, or rebound into partially open positions when they were closed, or make loud noises when they were closed, as they hit the backs of the inside of cabinets. Solutions to these problems lay in the design of better drawer slides. More generally, as one patent application put it, a cabinet had to be designed to "prevent sagging of the drawers."[77]

In the early decades of the twentieth century the office equipment industry experimented with a number of techniques to improve the "means for slideably supporting . . . drawer[s] within the cabinet."[78] By the second decade of the century, patented frictionless extension slides were used to hold the weight of the drawers when they were pulled out to full length. After a period of trial and error, most manufacturers had settled on a design that consisted of a suspension-type drawer slide that was equipped with four large and one small die-cast rollers on steel bearings.[79] Why? In the trademarked words of office equipment company Steelcase, "Nothing rolls like a ball."[80]

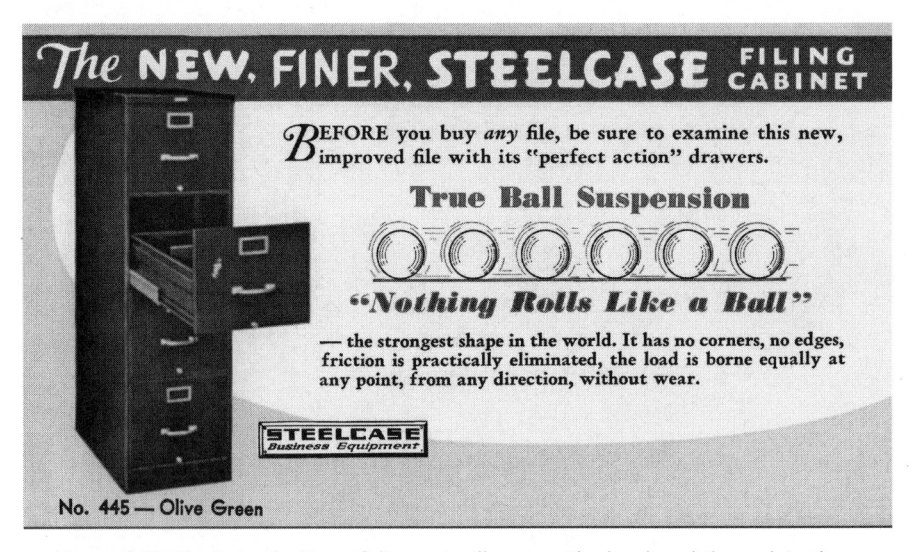

Figure 2.16. The introduction of die-cast rollers was the key breakthrough in the design of drawer slides. In this advertisement Steelcase used the slogan "Nothing rolls like a ball" to capture the efficiency of the company's drawer slide. Courtesy of Steelcase Corporate Archives, Grand Rapids, Michigan.

Access: Horizontal Filing Cabinets

To facilitate efficient access to the contents of a vertical filing cabinet, the drawer fronts had slots into which labels could be inserted to indicate which part of a numerical or alphabetical classification system each drawer contained. In another style of cabinet the sizes and shapes of drawers were used to initially identify the types of papers stored. In contrast to a vertical filing cabinet, the shell of a horizontal filing cabinet was designed to house different-sized drawers to accommodate the variety of paper records and documents used in an office. By their very shapes, the drawers marked the locations of specific types of paper, such as checks or index cards. In these cabinets, similar drawers were arranged side by side, usually either two or four drawers across, with the cabinet shell measuring 33–36 inches wide and 16–24 inches deep. Because the drawers were not stacked on top of each other, this was labeled a horizontal filing cabinet.

Companies promoted their horizontal cabinets by linking efficiency, productivity, and modularity. At the end of the 1890s, a Globe-Wernicke catalog noted: "Business is progressive, expansive. The modern cabinet must be progressive and expansive."[81] More than a decade later, rival company the Library Bureau explained the horizontal unit idea in more detail: "Each unit fits any other unit. Any single unit, with base and top or cornice, forms a complete cabinet. Any number or combination of units may be brought together into one cabinet." This cabinet would contain the files of a business, accommodated to the volume of paper and the size of the office. The Library Bureau continued its enthusiastic promotion, even suggesting stacking horizontal units vertically: "When floor space is limited it may be kept one unit wide and extend to the ceiling. Or it may be kept low in height and extend around the room. Or it may be arranged in various heights to fit the position of windows, doors, etc."[82]

The integrity of the filing cabinet spoke to concerns about protection and retrieval. The filing cabinet was designed to protect its contents from fire, dirt, animals, and theft. However, protection from unwanted visitors had to be balanced with the need to ensure ease of access to the contents for designated users; integrity had to guarantee access as well as protection. The integrity of the cabinet, its wholeness, was important not just for access but for access

Figure 2.17. This image from a Wabash Cabinet catalog shows a horizontal filing cabinet. Manufactured for smaller offices, its shell was designed to house different-sized drawers that could be arranged to fit the paper storage needs of specific offices. Wabash Cabinet Co., *Wabash Filing Cabinets* (1917), Trade Catalog Collection, Hagley Museum and Library, Wilmington, Delaware.

and growth; as John Durham Peters notes, "No container works best when completely full."[83] A storage technology that emerged from a business imagination entranced by efficiency, the filing cabinet had to anticipate that a business would grow as it became more efficient and productive. Storage continued to mark a place for storing. However, the enclosure it marked was shaped by a belief in provisional permanence. As one filing cabinet manufacturer succinctly put it, the modern office needed a cabinet that "is always finished but never complete."[84]

The demand that storage accommodate expansion, flexibility, and the movement of its contents generated its own version of the attachment anxiety that loose-leaf ledgers produced. Steel filing cabinet manufacturers sought to alleviate this anxiety through the design of drawers and cabinets. Drawers had to be easy to open, but the position of a drawer in a cabinet also marked an initial partition that pointed a user to the correct drawer. Demands for access shaped the interiors of the drawers as well. Within drawers, paper-based partitions further divided the interior space of a cabinet to make the classification system visible at a glance, to ensure that any given piece of paper could be located quickly.

Chapter 3

CABINET LOGIC
Efficiency through Partitions

The vertical filing cabinet offered a designated and discrete storage space. The cabinet's integrity promised continuity. Office clerks could be confident (according to promotional literature) in the guarantee that they could easily access and retrieve papers. Such a guarantee depended not only on the structural integrity of the cabinet but also on the cabinet's interior. The filing cabinet's contribution to the smooth flow of work in an office depended primarily on paper being stored on its long edge. Moving inside the filing cabinet, this chapter focuses on the connection between the physical support required to store paper vertically and the logic of segmented storage.

As I have already noted, William Leffingwell championed the vertical as a principle of efficiency, celebrating the filing cabinet as an illustration of his belief that efficiency demanded recognition that "there is almost always more room for growth in a vertical than in a horizontal direction."[1] In his 1925 book on scientific office management, Leffingwell noted the verticality of the filing cabinet as a structure of stacked drawers, but he was most enthusiastic about how it stored paper. This object of "great and peculiar importance" maximized vertical growth because it enabled paper to be stored "vertically" on its long edge.[2]

Leffingwell understood a filing cabinet drawer as a space-saving technology. He claimed that a filing cabinet drawer held ten times more paper than a flat drawer file or box file, and that it was able to do this because it took advantage of the vertical: "One hundred letters, laid flat, one letter high will occupy a space of about 84 square feet; if laid all in one pile they will occupy an area of 93½ square inches, while if set on the 11-inch end, they will occupy about 6 square inches."[3] According to Leffingwell, utilizing the height of a room, a stack of vertical filing cabinets would continue to store more paper with a smaller footprint than a pile of papers stacked to the ceiling. He also noted that storing paper on its long edge made it easier for workers to locate and retrieve individual papers: "In addition, letters filed on end, not only occupy less space, but make the problem of location much more simple and the operations of handling far less laborious."[4] Simply put, storing unbound paper vertically on its long edge made individual sheets easier to retrieve; verticality again provided a way to avoid congestion and encourage better work flow in the office.

For Leffingwell, the retrieval of paper involved locating it and then extracting it by hand from the larger collection. While the latter was particularly important to someone who regularly touted himself as a disciple of Frederick Taylor and his "scientific" systems for managing labor, both these aspects of retrieval were technologically intertwined inside the filing cabinet. The technologies used to keep paper on its long edge were also integral to the marking of designated locations for papers. The paper inside the cabinet and the classification system used to give individual papers specific locations both stood at attention.

The shape and arrangement of the vertical filing cabinet's interior, its "cabinet logic," made the cabinet a novel storage technology in the office. Cabinet logic involves the creation of interior compartments to organize storage space according to classification and indexing systems. It connects the filing cabinet to a larger history of storage technologies; this is the history of "orderly furniture" and "intellectual furnishings," most of which celebrants like Leffingwell ignored.[5] In her insightful analysis of boxes as sites of order and system, Anke te Heesen foregrounds an important aspect of cabinet logic. She argues that in eighteenth-century merchandise and natural cabinets, "conceptual viewing and the boxed object meet within their ordered categories."[6] The creation of specific places for objects in a cabinet according to a classificatory system gives each an identity and value different from what it has when viewed in isolation.

A historically specific cabinet logic, used in the name of efficiency, determined the interior of the filing cabinet. Partitions made from paper, not wood, divided storage space to create rigorous order; these partitions took the form of tabbed manila folders separated by tabbed guide cards. This iteration of the logic dispensed with a separate index to make paper discoverable by utilizing the "very organization of the material and its location" with the "vertical guides serving as locating medium."[7] Elimination of an index was signaled in filing literature by the terms "direct alphabet index" and "automatic index"; this also resulted in the blurring of any difference between classification and indexing in promotional literature.[8] In the filing cabinet, classification solved a temporal problem as well as the spatial problem of organization. Without the need to consult a separate index, a clerk grouped papers together on their edge behind tabs labeled with classifications, so any given paper could be found quickly.

As te Heesen argues, partitioned boxes and cabinets allow a "specific field of activity to arise between hand, thing, and box."[9] In the case of the eighteenth-century curiosity cabinet, the removal and return of objects introduced a haptic dimension to the significance of the cabinet and the knowledge it created for its contents: "direct grasping of an object" was necessary for the observer to ascertain its character and quality.[10] In contrast, with paper as its object of storage, a filing cabinet shaped a different encounter with hands (which I discuss in chapter 5). In contrast to earlier material iterations of cabinet logic, the importance granted to efficiency prioritized handling solely as an action to ensure the circulation and storage of the information the filing cabinet contained.

Within the filing cabinet, the partitioning of paper, the need for divisions, adhered to the increasingly dominant late nineteenth-century articulation of efficiency; it was mobilized to save time and space. This version of cabinet logic prioritized speed to such a degree that early twentieth-century office management advocates struggled to position the vertical filing cabinet in relation to what they considered the commonsense understanding of storage as a sedentary state. As previously noted, many office manual authors believed "any type of storage system is a passive agent in business."[11] If, as efficiency advocates believed, the office had become a place of growth and progress, of efficiency and movement, then storage had

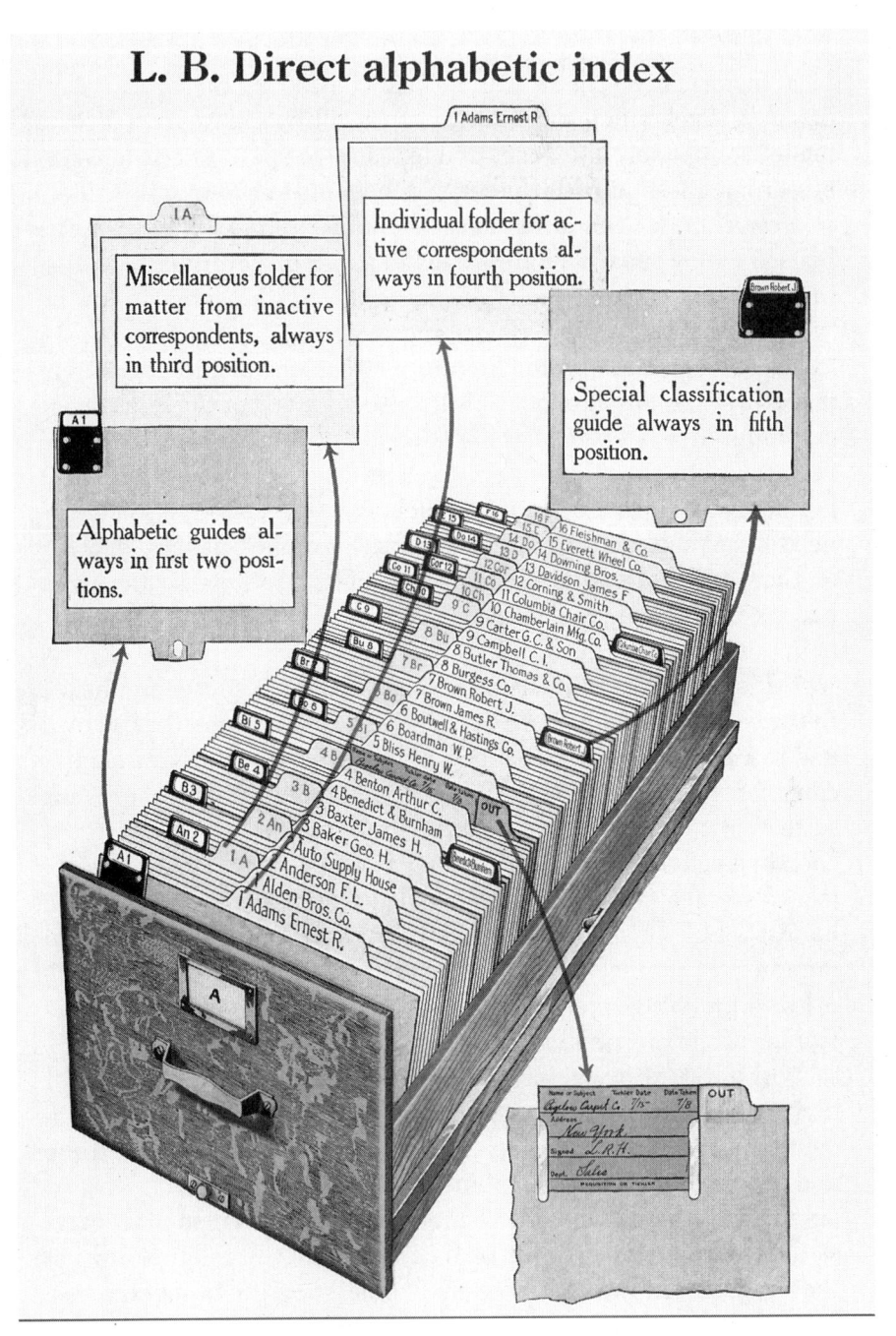

Figure 3.1. This catalog page shows one of the Library Bureau's specialized indexes intended to make specific papers easy to file and find among a large collection. An early selling point of filing cabinets was the inclusion of the index as part of the storage rather than as a separate registry or card index. Author collection.

Figure 3.2. This Library Bureau pamphlet cover emphasizes the importance of accessibility to vertical filing. To be accessible, files had to be visible; they had to stand on their long edge. Drawers, folders, and guide cards were designed to ensure that papers stood at attention, with tabs visible and folders taking up as little space as possible in a drawer. Courtesy of the Herkimer County Historical Society, New York.

to adapt. Paper should no longer be hidden in drawers or pigeonholes or lying flat in piles, "dead"; it needed to be alive and standing at attention. From this perspective, it was problematic to call filing "storage" because paper in a vertical filing cabinet was "not simply in dead storage, but in such fashion that any one of many thousands of letters and documents and records can be located instantly when needed again."[12] Vertical filing brought paper to life because it made classification a strategy for retrieval.[13] By prioritizing the speed in which papers could be found, classification became a temporal concern, not just a spatial practice for organizing knowledge according to categories.

In this chapter, I use the concept of cabinet logic to get under the hood of the filing cabinet to understand how it works as a designated location for particular papers. The historical specificity of this logic is part of my broader argument that storage is not neutral. Particular ideas and practices shape storage technologies; those technologies shape how people encounter stored objects. By identifying the vertical as a principle of efficiency, Leffingwell suggested the ideas that shaped the filing cabinet as a technology that defined storage as a problem of retrieval. To separate and locate particular papers, the filing cabinet used manila folders and guide cards. These not only identified the paper as information but also supported it, making the information visible and accessible. However, while they helped paper stand on its long edge, a specific technology was needed to keep paper in a vertical position—the follower block or compressor.

Vertical Paper: Compressors

> CHAPIN: Wouldn't the question come up what "vertical" means? Does a man know what "vertical" means?
>
> DAVIDSON: He will as soon as you call it to his attention and explain it to him. He does not need to consult a dictionary. . . . It seems to me as distinct and descriptive title as any we could use. I do not know who coined the application but it seems to be particularly good and expressive.[14]

This conversation occurred during a Library Bureau salesmen's meeting in Boston in 1899. The participants were discussing the merits of the vertical filing cabinet. For more than six years, the company's main office had largely ignored its Chicago office's invention. However, having now seen its value, the company president wanted to make sure that customers would as well. His faith that "a man" was capable of knowing what "vertical" meant when applied to filing and the storage of paper was well-founded. The label "vertical filing" stuck; seven years later, a patent applicant noted it was "now well known."[15] Writers who felt the need to define vertical filing stressed that it allowed papers to be placed "on edge" or "upright" in drawers.[16] The importance granted to having paper standing up on its long edge necessitated the invention of ways to keep the paper in that position. What did verticality require?

Before hanging folders became pervasive in the middle of the twentieth century, the primary technology used to keep paper vertical in file drawers was a specifically designed piece of wood or metal. In addition to keeping papers upright, it was intended to save space by keeping papers compact; patents used "confine" as a descriptor of choice.[17] The technology had two names: compressor, which drew directly from the process; and follower block or

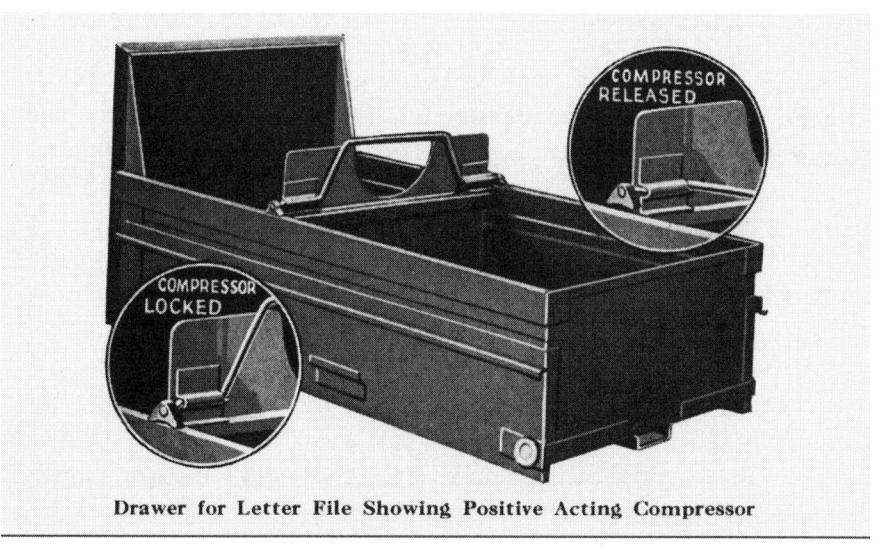

Drawer for Letter File Showing Positive Acting Compressor

Figure 3.3. This General Fireproofing catalog page features the main problems people encountered with a compressor. To function effectively, a compressor needed to support papers but also had to be easy to move and easy to lock and unlock. Author collection.

follower board, which came from the object itself. The latter name also noted the object's placement in the drawer, where it followed the papers.

Regardless of its name, the piece of wood or metal had two immediate goals. In addition to "preventing sagging or slumping in drawers," it needed to maintain a "compact and orderly arrangement."[18] As the latter quote indicates, the storage of paper vertically was intended to serve the purposes of confinement, classification, and circulation. A compressor showed the connection between verticality and cabinet logic; if late nineteenth-century cabinet logic put storage in the service of timely retrieval, the filing cabinet's significance lay in how the structure of the cabinet facilitated the accessibility and circulation of paper.

A compressor was called such because it pressed separate things together. As the entry for "compressed" in the 1891 *Oxford English Dictionary* defined the term: "pressed together closely, so as to occupy small space; pressed into a smaller volume and denser composition than the ordinary; condensed." The compressor in a filing cabinet did not reduce the size of the papers, but it managed storage so that a collection of papers occupied as little space as possible. The importance of a compressor to the spatial organization of paper highlights that compression is a matter of matter occupying space (as a comparison with "data compression" suggests, what differs historically is the matter and manner in which matter occupies space). To maximize capacity, vertical filing cabinets reoriented paper storage from the horizontal to the vertical; to achieve that, a compressor performed the role of gravity in the piling of paper. If sheets of paper are piled one on top of the other, gravity reduces the gaps between sheets of paper to maximize the number of papers that can be piled in a given space. Although a compressor took on the work of gravity, a filing cabinet had to do more than create spatial efficiency; it also had to create temporal efficiency, so that an individual folder or paper could be easily accessed.

In the design of early compressors, office equipment companies encountered problems in their attempts to address the dual concerns of capacity and access. A compressor needed to be both stable and flexible. If it was not, papers could be damaged and filing could also become more time-consuming. To be effective, the device needed to be securely held in place so it could support papers, but also easy to adjust so papers could be added to or removed from a drawer. While one origin story has the vertical filing cabinet directly inspired by index card or card catalog cabinets, the type of follower block used in those drawers could go only so far as a model.[19] The needs of the filing cabinet, with its more regular insertion and removal of papers and the greater weight of papers relative to cards, demanded a follower block with a different degree of compression.

One patent applicant described a follower block that, once placed in a drawer, would be "easily and quickly operated from one position of adjustment to another"; it would also "automatically lock itself in a manner to effectually resist movement except at the will of the operator."[20] Echoing the design of the library card catalog, the compression mechanism was initially attached to a guide rod that ran along the bottom of the drawer. One company celebrated "the metal-on-metal operation of spring and track" as it promoted the ease with which its compressor could be used.[21] A file clerk could release a lever attached to a spring that ran from the top of the block down to the rod, allowing the compressor to be moved backward or forward; the promise was that a clerk could do this with ease.

However, despite the "will of the operator," early compressors often could not be moved easily from their original positions, or they would catch on the sides or bases of the drawers. For some users, it seemed the manufacturer's priority was to keep papers upright at the expense of access. Championing an early but failed attempt at hanging folders, one company catalog argued that the logic behind a compressor was that "unless *tightly* 'wedged' [papers] will crumple and fall." This created the problem that if "tightly wedged they are almost inaccessible" unless a person used "both hands and considerable muscular effort."[22] If it was only about "confinement," storage was prioritized over retrieval.

These complaints provided motivation for a number of patent applicants as the office equipment industry sought different solutions to the problem of compressing papers. A consensus emerged that the guide rod was problematic. A commonly used solution involved the insertion of "channels" in the sides or bottom of the drawer. The compressor moved along these welded channels on rollers attached to the edge of the board; a locking mechanism was usually still attached to the rod.[23] An alternative was a guide path in the bottom of the drawer with which the "toe" of a locking lever would engage.[24] In this design, movement was usually initiated through the tilting of the board forward or backward to unlock it. Or, as it was described in an Art Metal office equipment catalog: "Pressure of the finger releases the follower, which glides smoothly in a galvanized channel in the drawer bottom until brought to position—removing your finger instantly locks the compressor. The compressor is compact in construction with just the right slope for easy reference to papers and allows maximum filing space."[25] While not explicitly gendered in this advertising copy, the finger represents the ease with which a filing cabinet could be operated, as Shaw-Walker suggested in its illustration of a young girl opening a file drawer by pulling on a single piece of thread (Figure I.6).[26]

By the 1930s, another solution had emerged to keep files in an

Then—In Goes this "Press the Button" Follower Block

Why does it slide along so smoothly—no catching or jerking? Because it doesn't depend on the *guide-rod*—it runs on rods of its own—two of them. Something brand new—and much needed. Adjusts easily. Locks instantly. One finger does the whole thing. This is one more example of Shaw-Walker ingenuity.

Figure 3.4. As compressor design improved, companies emphasized how easy it was to operate newer compressors. This page from a Shaw-Walker catalog illustrates the combination of push-button and one-finger operation that companies often employed to represent ease of use in an age of machines. Shaw-Walker, *Filing Cabinets in Wood* (1919), Trade Catalog Collection, Hagley Museum and Library, Wilmington, Delaware.

"upright and orderly position."[27] Office equipment companies marketed removable enameled steel plates that fit into slots or grooves in the bottom of a filing cabinet drawer; these were given various names, including "sway block" and "flexiblock."[28] A drawer would usually have five or six of the "blocks" to compress folders and maximize capacity. Office equipment company Shaw-Walker marketed its version of the enamel plate as the Wobble Block, claiming, "It flops back and forth (hence its name)."[29] In moving back and forth, it released (decompressed) folders to create a V-shaped opening at the desired point of reference, reducing the effort involved in retrieving a file. This would happen when a clerk's fingers touched a folder, an action that forced the material in front of it to move forward; the resulting weight would then tilt the nearest plate forward, while the material behind it moved backward to pressure another plate to move backward. This created the V that allowed

The
WOBBLE BLOCK
Transforms Drawers into Filing "Machines"

Figure 3.5. In the 1920s, prior to hanging folders, up to half a dozen removable metal plates could be placed in drawers to keep papers standing in place. As this image from a 1940 office equipment catalog illustrates, Shaw-Walker called its version the Wobble Block. It stopped papers from moving and could be easily moved. Courtesy of Grand Rapids History and Special Collection, Archives, Grand Rapids Public Library, Grand Rapids, Michigan.

easy retrieval of the folder from the previously tightly compressed drawer.

The Wobble Block and its kin were attempts to create compression and decompression more effectively in response to the realization that extreme spatial compression inhibited access by making it difficult for fingers to extract folders. Either papers had to be stored with reduced compression or compressors had to decompress effectively, creating the sought-after V opening, so that individual folders could be easily retrieved. Therefore, any space the compressor created by squeezing papers together had to allow for the temporary expansion required to enable a worker to access papers. The temporary expansion cut into the space the compressor had created by pressing papers together. As a result, because of the expansion that decompression required, the space a compressor created in the drawer could never be completely filled.

The support provided by plates dispersed through the drawer also promised a better way to keep paper vertical irrespective of its size or format. The original compressors were now presented as flawed because they were not efficient at compressing different sizes of documents evenly, and they did not work well in drawers that contained both papers and books. Ignoring claims they had made for decades, filing equipment companies changed tack to argue that instead of preventing files from sagging and slumping, compressors created this problem by not supporting the long-edge verticality of different paper formats equally. Adding a number of metal sheets to a file drawer purportedly produced compartments such that "heavy books and papers can be easily filed in the same drawer" without papers becoming curled or damaged, and with the added benefit of not obstructing index tabs.[30]

The need to have tabs visible underscores how the compressor worked to facilitate access. When papers were compressed and standing tall, retrieval also depended on gaps, in the form of visible divisions. In this case, it was not a clerk's hands but her eyes that had to access the classificatory/indexing system. As Cincinnati-based Globe-Wernicke argued in a 1930s steel office equipment catalog, "Sound filing practice requires material in a drawer to be kept in an upright, orderly position," because "slumping files mean sluggish offices."[31]

Particularity: Tabs

To help their customers find specific papers, office equipment companies added tabs to folders and to the guide cards used to divide folders. These "sign-posts in the file" made visible the indexing system used to arrange the file drawer.[32] This enhanced the degree of particularity critical to the goal of rapid retrieval prioritized in this iteration of cabinet logic. These "indexing projections" provided spaces for labels identifying the contents of folders or divisions in a filing system.[33] Underscoring the prevalent use of alphabetical systems, one catalog stated that tabs made visible an arrangement "based on the telephone directory with which everyone is familiar."[34] The number of tabs used to label divisions (as opposed to particular content) depended on the number of guide cards in a drawer. Advertising and how-to literature did not offer uniform rules on how many guides should be used in a drawer, but some authors gave specific advice, recommending one guide for each inch of filing depth, for example, or fifteen folders per division. Decisions about the number of guides also needed to take into account the fact that guides "[took] up expensive filing space"; the more guides, the less space for files.[35]

To produce efficiency, tabs in a filing cabinet had to be clearly visible, so the eye could quickly lead the hand to the desired papers. Although tabs should ideally "always stick out in plain sight like a sore thumb," this was more easily stated than achieved in early vertical files.[36] Problems arose as tabs tended to become damaged, dirty, and difficult to read over time, as they were subjected to regular use. The problems associated with wear stemmed from clerks' having information at their fingertips, literally. The tabs absorbed moisture from "the hands of user[s] while being handled."[37] Frequent handling also caused tabs to become bent or curled over. Clerks exacerbated both these problems by regularly using tabs as handles when they removed folders from drawers.[38]

Therefore, in the first decades of the twentieth century manufacturers attempted to reinforce and protect tabs, particularly tabs on guide cards and more expensive folders. Celluloid was the material of choice for these tabs. It was initially applied to the fronts and backs of preprinted tabs to protect the index information recorded on them. On stronger (and more expensive) pasteboard guide cards, manufacturers soon replaced the celluloid-coated tab with a piece of celluloid folded over at the top edge of the card, forming a pocket in which a label could be inserted. These label-holding tabs tended to split across the top, however, and the cracks allowed dirt to collect on the written labels. The cracks also created the risk of injury to the "fingers of the user."[39]

In an attempt to reduce these problems, manufacturers began to enclose preprinted guide labels in solid pieces of celluloid that left no exposed edges. The initial adoption of metal-framed tabs repeated previous problems as the edges of the light tinned metal often broke off and inserts frequently worked their way out of the frames. By the 1920s, manufacturers had settled on celluloid tabs with slots in the back or side for removable labels for expensive folders and cheaper guides to accommodate expansion. Expensive

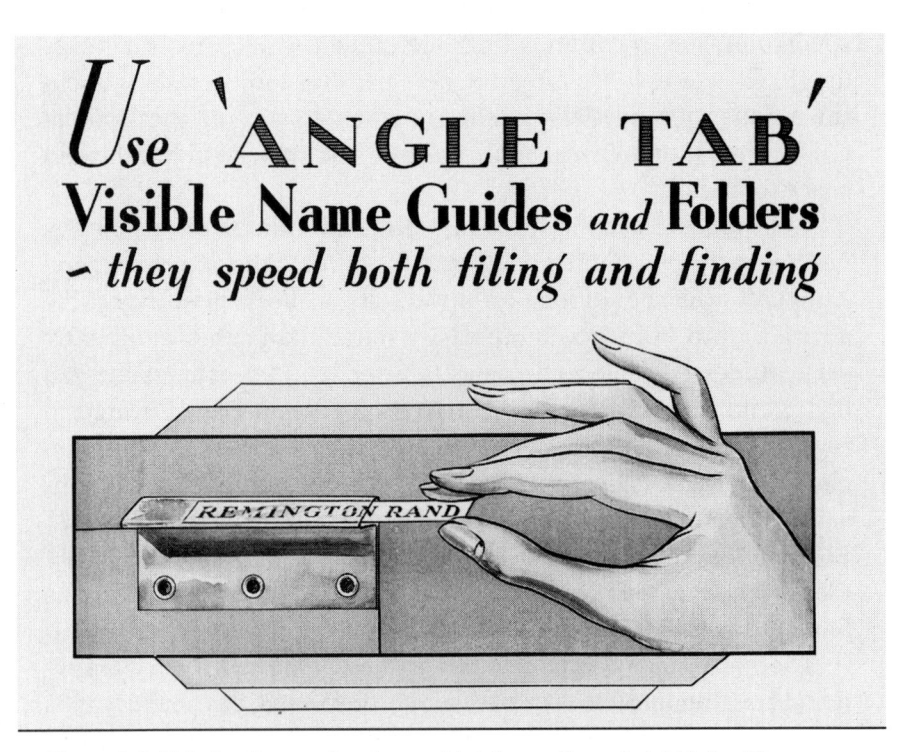

Figure 3.6. This Remington Rand pamphlet shows the celluloid tab with removable inserts that became the default form of tab for expensive guide cards and folders. These tabs made of nickeled metal were securely attached to guide cards. Courtesy of the Herkimer County Historical Society, New York.

guides used a heavier metal (nickeled) with rounded corners and edges and celluloid windows. A triangular piece of metal (a "shank") that extended farther down the card using two or three eyelets attached the tab to the guide.

Regardless of how tabs were made, position and color were deployed to try to ensure that they worked successfully in a file drawer. Initially, the basic low-cost folder produced by most manufacturers featured a back that was a half inch higher than the front. This created a tab with sufficient space for someone to write a few words to identify the folder's contents. However, in the name of precision and efficiency, file equipment companies changed their approach to placing tabs. The back of a folder or guide was envisioned as a horizontal plane that could be divided into five or six equal lengths. Tabs were then cut from or attached along the top in one or more of those positions. These tabs were intended to be complementary, so the tabs on guides and folders would not obscure other tabs when the items were placed in a drawer. A common arrangement had guides tabbed at the left or first cut and folders tabbed from the right, starting at the last cut; on multitabbed folders the tabs would go into the third or fourth cut.

In the 1920s, advertisements emphasized the positioning of tabs in terms of degrees, specifically 45 degrees, "the natural reading angle," according to one advertisement.[40] This was the angle at which a tab would rest atop a folder. As they did with all aspects of filing, office equipment companies linked the angled tab to the ideas of efficiency they used to sell office equipment. Invoking scientific management's focus on the worker's body, a Globe-Wernicke advertisement claimed, "There is no stooping or pushing content of drawers about in order to read labels." Positioned at a 45-degree angle, its "tabs look you straight in the eye."[41] Ease of visibility meant that the clerk did not have to stoop or bend the tab to read it; the tab possessed agency to look at the clerk. Less handling meant a tab would not be soiled: "Always clear, always clean, it will endow your files with an efficiency never before attained."[42]

Guide cards were also designed to help keep folders upright so that tabs were visible. Made from heavy manila paper or two different weights of pasteboard, they were sturdier than correspondence paper and ordinary manila folders. Promotional materials exuberantly proclaimed the possibilities of guide cards for supporting filed paper, celebrating how they offered "exceptional strength—sufficient to hold the heavily crowded folders at all times upright in the file."[43] Without sufficient guides, customers were told, it would

Tab Cut Folders
(Plain)

One-Fifth Cut Folders
9½" High Overall

	Letter Size	Cap Size	Invoice Size
Medium weight	No. 415	No. 515	....
Heavy weight	No. 425	No. 525	No. 625
Extra heavy weight	No. 435	No. 535	....

One-Third Cut Folders
9½" High Overall

	Letter Size	Cap Size	Invoice Size
Medium weight	No. 413	No. 513	....
Heavy weight	No. 423	No. 523	No. 623
Extra heavy weight	No. 433	No. 533	....

One-Half Cut Folders
Right and Left Positions
9½" High Overall

	Letter Size	Cap Size	Invoice Size
Medium weight	No. 412	No. 512	....
Heavy weight	No. 422	No. 522	No. 622
Extra heavy weight	No. 432	No. 532	....

Figure 3.7. This General Fireproofing catalog page shows examples of different styles of tabbed manila folders. These ranged from one tab to five tabs. Multiple tabs were intended to be complementary to ensure that the tabs on following folders would be visible. Author collection.

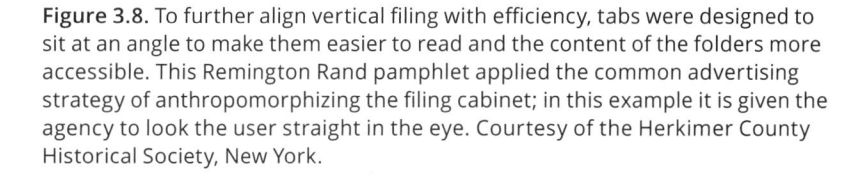

Figure 3.8. To further align vertical filing with efficiency, tabs were designed to sit at an angle to make them easier to read and the content of the folders more accessible. This Remington Rand pamphlet applied the common advertising strategy of anthropomorphizing the filing cabinet; in this example it is given the agency to look the user straight in the eye. Courtesy of the Herkimer County Historical Society, New York.

be "push and pull for the file clerk all day long."[44] More expensive pasteboard guides usually had metal eyelets in the bottom center so they could be threaded onto a rod that would keep them in place and upright; some companies offered guides with three rods (one at the bottom and one on each side).[45] The rod attached to the bottom of the guide was used to keep papers and folders from slipping down into the base of the drawer.

Gathering: Folders

Inside a vertical filing cabinet, papers were stored in manila folders. As a file, the manila folder functioned to gather and partition groups of papers. It provided tangible boundaries within which all papers that addressed the same "matter" could be placed. Or, put another way, as the object that was arranged numerically or alphabetically, the manila folder became the basic unit of vertical filing. Used within a vertical filing cabinet, it was shaped by principles of verticality and integrity, which in turn were shaped by ideas of efficiency. A tabbed manila folder represented a new storage format for organizing things, notably paper, information, and files.

The gathering function of the manila folder locates it in the history of files. As an object that demarcates space for papers and as an entity that can be acted on, the "file" has collected paper in many different ways and been stored in a variety of places. Its name in English comes from techniques to gather paper introduced throughout Europe at the beginning of the fifteenth century. Papers strung on wire or string became known as "files" care of the Latin word *filum*, meaning string or thread, via the French *filer*, to spin and thread. Usually the papers that constituted files were secured together by string, rope, or leather straps. Therefore, the file is a technology of gathering that brings a collection of papers together as a unit to be placed somewhere.[46]

Within institutions, circulation became an additional feature of the file. This produced the administrative file as the dominant file form.[47] It gathered different types of paper records on a similar topic, often conceptualized as a "case" to be circulated around an office.[48] An administrative file puts separate documents into a "co-documentary context," where, dependent on each other, they create a case; the event or issue that initiates a file reconfigures the purpose and focus of the papers.[49] A case file usually has some form of binding to ensure that the order of the papers is maintained. Order is also important to its path in an institution, as it is designed to move from one designated official to another. Those officials often record the path of the file on its cover; their initials are viewed as evidence that they read the file. The comments and papers added during the circulation of a file produce a distinct mode of administrative writing. In its regulated movement between officials, a case file produces the authority of the institution; this does not automatically represent existing hierarchies, as officials can shape the movement of a case file for their own advantage.[50]

The filing cabinet and manila folder are part of this history but represent a change in need and emphasis. The manila folder is a file format that reverts to gathering or storage as its primary purpose: holding unbound papers in a drawer. It is designed to highlight the place of papers in a classification/index system. This is intended to safeguard access, to ensure the folder or papers in it can be retrieved. Although the papers in a manila folder and the folder itself can be circulated, that is a secondary function, particularly when compared with the function of an administrative file, the circulation of which is critical to decision making and to the exercise of institutional authority.

Put simply, a manila folder is a less rigorous form of file. Although it provides a place for papers in a drawer, it is not designed to maintain order and produce authority within an institution. The flexibility a manila folder offers expresses the needs of a commercial office shaped by the demands of efficiency. Containing unbound paper, it emphasizes the particular, the granular, and prioritizes the opportunity to extract, to recombine discrete parts in new ways if necessary. A tabbed manila folder is critical to the joining of efficiency and the conception of information as a discrete unit. Unbound paper stored in a manila folder provided a new and important conceptual gateway for understanding information as something different from knowledge in its particularity and instrumentality; it gave information presence.

That the manila folder surrounded papers, enveloped them, was its central contribution to the innovation of vertical filing that Leffingwell and others highlighted: when enclosed in folders, loose papers were easier to store and retrieve without damage to the

Straight-edge folder

Individual tab folder

Pressboard expansion folder—tab cut

Pressboard expansion folder
Removable label tab

Figure 3.9. The manila folder became the basic organizing unit of the vertical filing cabinet. This page from one of the Library Bureau's vertical filing catalogs illustrated the different styles, shapes, colors, and grades of paper available in the company's line of folders—they were not always manila. Author collection.

papers or to the integrity of the storage system.[51] In 1898, during the early days of vertical filing cabinets, when not all initial buyers used folders, the president of the Library Bureau, a former folder skeptic, announced to a salesmen's meeting, "It is a distinct service to have the folders, you can pick the things right out. Things can't get out of place."[52] In explaining how to use a folder outside a drawer, how-to-file authors commonly appealed to their readers' familiarity with books. Although the folder's value came from the fact that it stored unbound paper, filing instructors initially referred to the book as a model for reading the loose sheets of paper in a folder. Clerks were encouraged to place papers in a file folder with their left edges at the bottom of the folder and their tops toward the folder's left end. This meant that when a folder was taken out of a drawer and placed flat on a table, "you refer to the contents just as you would turn the leaves of a book, except that the papers are not fastened."[53]

The folder's value as an enclosure came from its capacity to contain different sizes of papers (this marked a significant difference from the card catalog, which stored only one size). A series of standard-size folders turned an assortment of paper sizes into a file. While papers of different sizes could fit in a folder, folders also came in different sizes for the three main types of filing cabinets offered by office equipment companies: correspondence, legal, and invoice. These cabinets, and the folders intended for them, differed slightly in size even within given types, because in the early twentieth century no industry-wide standards for paper sizes existed in the United States. One expert writing in the late 1920s claimed that American offices used twenty-four different sizes of paper for record keeping, but in that decade support began to coalesce in American business for two sizes of paper for correspondence: 8.5 by 11 inches and 8 by 10.5 inches (the two sizes were advocated by separate government committees).[54] In contrast, in 1922 Germany introduced a standard paper format based on the metric system, and the size designated as A4 became the standard for office work; an A4 sheet measures 210 by 297 millimeters, or 8.27 by 11.7 inches. The person officially credited with inventing the A4 paper format anticipated the impact the standardization would have on office furniture and started a company that produced binders, folders, filing cabinets, and bookshelves sized for the new format.[55] In the United States, the lack of standards in paper sizes during the interwar period meant that office managers needed to check with their file clerks before purchasing folders.[56]

The manila folder's function of enveloping loose paper brought with it the same kinds of concerns about attachment and integrity that accompanied other unbound paper storage technologies, such as the loose-leaf binder. In response, filing equipment companies offered vertical file folders with metal fasteners designed to keep papers together, making the folder into a technology of attachment. A popular version was the "tang folder," which had two metal prongs riveted near the top of the inside back; papers in the folder were threaded onto the prongs. Other alternatives used to fasten papers to one another within a folder involved paste, pins, or clips. Any mode of attachment created bulk, however, and experienced file clerks warned against crowding drawers with too many folders using fasteners.[57]

The space a folder occupied in a drawer was of minimal concern in law offices. Tang folders and file envelopes with flaps and string closures were preferred in these settings, as they ensured that papers would not fall out when the folders were carried from place to place within and outside the office. This attention to securing the placement of papers acknowledged the importance attached to order and authority in the history of legal files. When "envelope" folders were used, offices tended to use different folders for the various aspects of a case; as one writer noted of these folders in 1918, "In carrying suit papers to court in keeping them apart from one another while in the hands of the various members of the firm they are invaluable."[58] Folders with prongs inside on both front and back were used to separate correspondence and legal documents. For example, a pronged folder could have drafts and a final copy of a will on one side and correspondence related to the will on the other.[59] On both sides of the folder, fasteners not only kept the papers secured but also kept them in chronological order, with the most recent documents placed at the front. Another technique was to keep the correspondence and legal documents for a given case in separate folders (fastened to the folders' backs). Law firms' ongoing investment in completeness and attachment extended to their treatment of folders when a case was closed: typically, at that point copies of all briefs, with the transcript of record, were removed from files and bound; the remainder of the documents and papers were kept as a file and stored, often off-site.

In businesses other than legal offices, concerns about the integrity of manila folders usually focused on protection. For a folder to protect papers it had to be made of a material stiff enough to stand up to the wear and tear of use, especially regular opening and closing.

Binder folders

Grade and weight	Correspondence size Cat. no.
De Luxe, Manila, with individual tab	5008F

In addition to the above folders which are carried in stock, any of the folders in the foregoing illustrations may be fitted with fasteners, making binder folder.

L. B. Binder folder

Crystaloid Folders

9⅞ in. high (inc. ⅜ in. tab.), 11¾ in. wide. Made of De Luxe, 18 pt. extra heavy Manila.

Description	Correspondence size Cat. no.
Plain label, 3rds asst.	5093K
Plain label, 5ths asst.	5095K
Individual tab	5098K

Crystaloid folder

Leatherette pocket

Description	Correspondence size Cat. no.	Legal size Cat. no.
1¾ in. expansion	507A	527A

Leatherette folders—straight edge

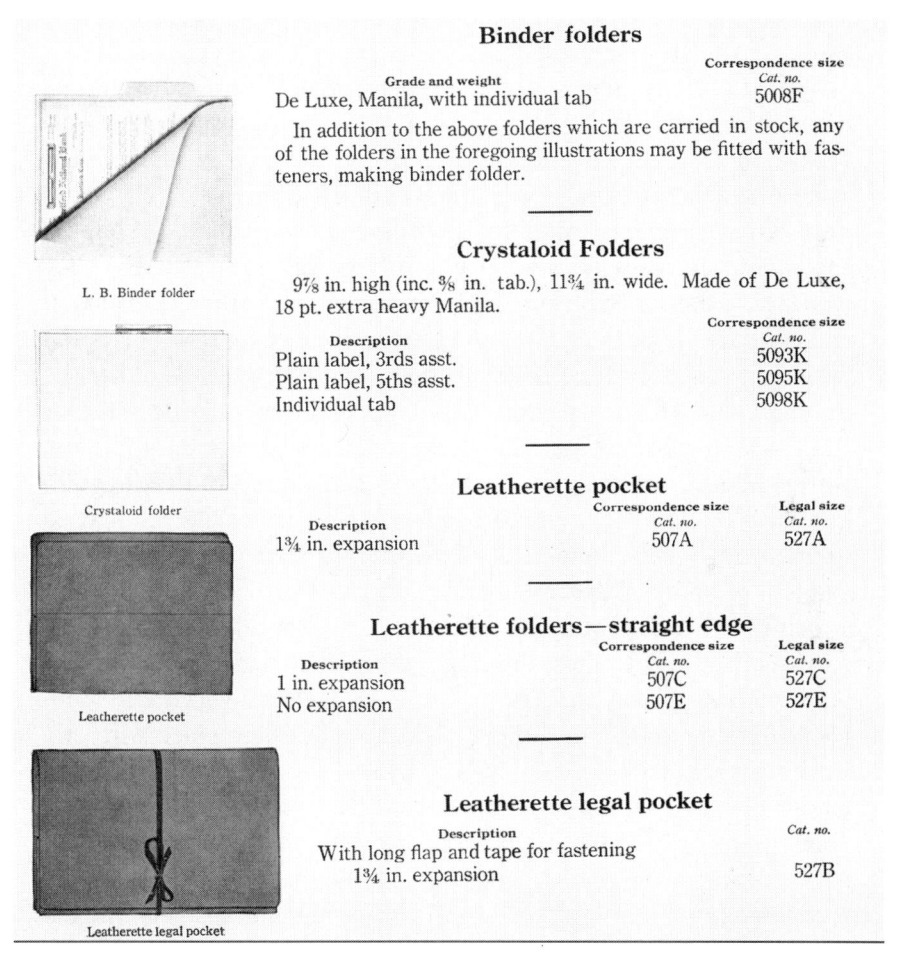

Description	Correspondence size Cat. no.	Legal size Cat. no.
1 in. expansion	507C	527C
No expansion	507E	527E

Leatherette pocket

Leatherette legal pocket

Description	Cat. no.
With long flap and tape for fastening 1¾ in. expansion	527B

Leatherette legal pocket

Figure 3.10. The filing cabinet was promoted as giving loose paper the integrity that a bound book gave to pages, and folders could be made into "pockets" or fitted with fasteners to make them equivalent to binders. This style of folder, illustrated in this page from one of the Library Bureau's vertical filing catalogs, was particularly popular in offices where folders were regularly circulated outside cabinets, such as in legal firms. Author collection.

A basic folder made of a single sheet of manila paper quickly became common across the office equipment industry. Manila paper was thicker than paper made from wood pulp. It was produced from abaca fiber, which came from a species of banana unique to the Philippines; the material was known colloquially as Manila hemp because it had arrived in the northeastern United States by way of "grass rope" on Filipino ships. A product of the circulation of objects and use structured by capitalism and empire, manila paper was patented in 1843.[60]

Manila folders were produced in three different thicknesses, measured in "points" ranging from seven to eleven; a point was equivalent to 1/1,000 of an inch. Folders of different point sizes were offered to accommodate the storage and retrieval roles that folders performed in offices, specifically, whether they contained papers that would be accessed regularly or occasionally and whether they would hold papers that were to be stored for short or long periods. It was accepted that, to support the weight of its contents when being taken out of a drawer, a folder needed to be constructed so that the manila paper's fibers or grain would run from top to bottom rather than side to side.[61] Some folders, while still called manila folders, were made from wood pulp. Sulfate was heralded as the key to high-quality file folders for regular use or long-term storage.[62] So-called extra-strong folders were made of pressboard—two pieces of 25-point cardboard bound gusset-style and reinforced with linen.

An important protection offered by the manila folder was its supposed ability to stop the bogeyman of curled paper. Curling paper was an initial problem in filing cabinets, as very early models frequently did not use folders; loose papers were placed directly into drawers, with guide cards marking the large divisions between categories. This sometimes resulted in paper "crumpling" or "buckling," or working up underneath the guides.[63] Into the 1920s some experts continued to suggest that folders could be eliminated if drawers contained papers of uniform and standard size; the example often given was order forms.[64]

However, folders were not guaranteed to protect papers and order in all circumstances. File clerks were warned about stuffing too many papers into a folder (with anywhere from fifty to one hundred pieces offered as a usable maximum). One classroom guide suggested that teachers define "congestion" as "an accumulation of papers in such volume that reference to any of them is difficult."[65] In practice, congestion could result in papers standing higher than the folder's edge, obscuring tabs and making it difficult to distinguish the divisions between folders. Other problems included trying to fit too many folders into a single drawer and placing one folder inside another. As was almost always the case, any misfiling was attributed to the worker, not to the equipment or the system.[66]

Filing equipment companies answered concerns about capacity and drawer space with folders that were made to accommodate the regular increase of content within specific categories.[67] Known as expansion or accordion files, they had additional folds at the bottom to permit them to expand about an inch while keeping the tab visible. Another option was a folder reinforced along the top.[68] This targeted an area susceptible to strain because, as noted previously, clerks often used the tops of folders as handles or bent the tops back as they moved folders forward while looking for other folders. Reinforcing only the top meant that folders could be more durable but not take up any additional space in the drawer; a heavy-stock folder could take up twice as much space as a regular folder.[69]

Jon Agar argues that paper files are "information technologies; they are material artifacts that store and organize information."[70] For Agar, who was one of the first scholars to argue for the need to look at the form of a paper file rather than its content, a focus on the materiality of the file offered another challenge to the belief that the contents of an administrative file provide historians with objective evidence. He suggests an approach that questions claims to objectivity through an analysis of files as material artifacts. This would involve establishing the relationship of files to filing systems and the systems of paperwork that precede their creation; he argues that the "'bureau' should be put back into studies of 'bureaucracy.'"[71] Agar's contention that such an approach shows the mechanization of government work raises epistemological questions shared by this book, albeit here the primary focus is shifted to business offices.

Although one reason for getting under the hood of a filing cabinet is to think about files as information formats, unlike Agar, I pair my interest in the materiality of the folder with an interest in the materiality of information. In exploring the materiality of information I argue that information as a category is a historically specific concept, and the emergence of the tabbed manila folder provides an object through which to understand the development (not the origin) of a distinctly modern conception of information as a thing that exists in the world, as something that is impersonal, discrete, and easily extracted.[72] In exploring this claim I am putting questions of storage at the foreground of the history of the filing cabinet, subsuming questions of organization within that focus, and, therefore, incorporating the file into the materiality of the filing cabinet.

To consider the relationship between storage and information is to position the critique of the file folder and the filing cabinet as part of a broader argument that storage is not neutral. To argue that verticality and integrity are principles of storage and to approach the interior structure of the filing cabinet as an example of cabinet logic is to foreground the values that shaped the filing cabinet. In the early twentieth century an encounter with a filing cabinet was an interaction with an emerging set of economic ideas. Filing is not a neutral act; it is shaped by efficiency, a distinct worldview that privileges a particular relationship to labor and time (it makes them things to be "saved"). Efficiency made cabinet logic an instantiation of granular certainty, the faith that breaking labor and time into small parts made them easier to control; information was something else that needed to be made granular. Therefore, to examine filing is to think about use as part of the materiality of the filing cabinet; it makes explicit the power dynamics that constituted the filing cabinet as an information technology and a storage technology. These are the arguments I develop in the remainder of this book.

PART II
FILING

Chapter 4 | Chapter 5 | Chapter 6 | Chapter 7

Chapter 4

GRANULAR CERTAINTY
Applying System to the Office

In 1900, A. W. Shaw (formerly of Shaw-Walker) launched a periodical titled *System*. It began as a twenty-page pamphlet with no page numbers, and within three years it was a ninety-six-page magazine with the subtitle *The Magazine of Business*. In 1903, Shaw founded a publishing company with *System* as its flagship; over the following decades it published many business-related books (including books central to the project of scientific office management). According to a former staff writer, by 1907 *System* had a circulation of more than 250,000 and more pages of advertising than any other magazine in the United States; even accounting for hyperbole, these claims suggest that *System* had significant currency in the business world of its day.[1]

System showed that the application of system in the office required a very particular kind of paperwork. Almost every other page of the magazine's early volumes carried an illustration of an index card, ruled with lines to store information in tabular form; on rare occasions, the image was of a ledger page rather than a card. These illustrations accompanied articles that explained how such cards could be used to organize records, correspondence, and information in a range of offices, including shipping, insurance, policing, dentistry, and real estate. In one article explaining how an agricultural implement manufacturer used index cards to keep track of patterns, the uncredited author stressed that "upon this card in the proper places, is recorded all the information about particular patterns. By means of proper guide cards these may be divided into classes as best adapted to your needs, such as machinery, current production etc. These cards should be of convenient size for keeping in a card cabinet (about 4 inches × 6 inches)."[2]

This "proper" placement of cards, and the information they contained, connects the celebration of system to the cabinet logic that increasingly structured paper storage in offices at the beginning of the twentieth century. As discussed in chapter 3, cabinet logic involves the creation of interior compartments to organize storage space according to classification and indexing systems. In the last decades of the nineteenth century, the partitioning of loose paper in file drawers emphasized the particularity of unbound paper. In this process, cabinet logic was articulated through dominant ideas of efficiency. The focus of this chapter is how that particularity was increasingly comprehended as information, as a thing that existed in the world distinct from knowledge, a thing that people could file and retrieve.[3]

Business publications such as *System* increasingly used the word *information* to refer to a fragment of knowledge that had a particularity and specificity. As the above description of the index card for patterns shows, the tabular division of a card provided boundaries within which the particularity that characterized "information" was created; in other contexts it might have been called "data," especially when the particularity took numerical form.[4] In the 1920s, a new type of card appeared in some larger American offices, as punch card machines were adapted to mechanize bookkeeping. They too were frequently described as a technology for storing and processing information.[5]

In this chapter I use the concept of granular certainty to explore how a file retrieved from a vertical cabinet shaped an encounter with paper, which in turn can be seen as an encounter with this particular conception of information. Granular certainty deploys a logic central to economically driven ideas of efficiency (and to the conception of information as a discrete unit). Within an economic context, the term *granular* signifies the belief that breaking things down into small parts to produce a high degree of detail or specificity will result in efficiency. *Certainty* indicates the conviction that

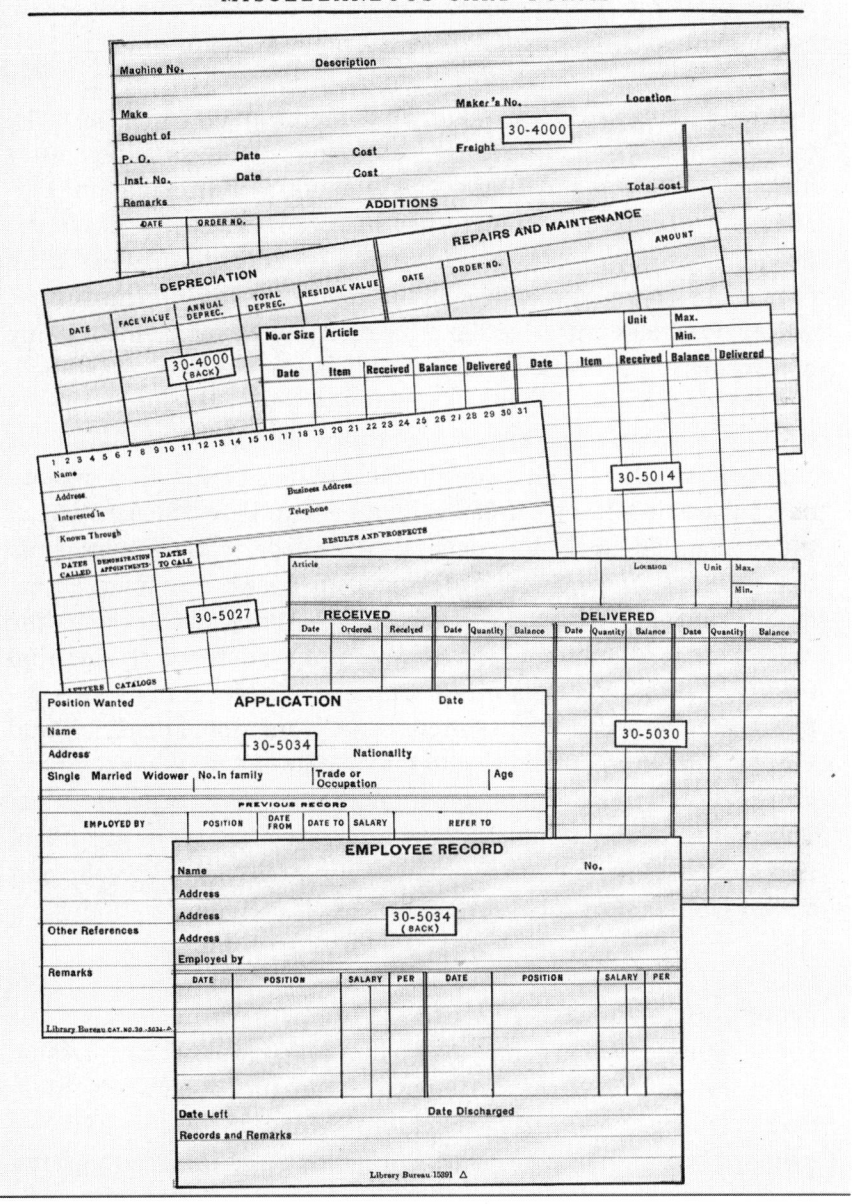

Figure 4.1. This Library Bureau catalog page shows some of the index cards the company offered. Index cards became a critical technology for recording and highlighting information across multiple aspects of a business. Library Bureau, *Card and Filing Supplies* (1920), Trade Catalog Collection, Hagley Museum and Library, Wilmington, Delaware.

greater specificity will reduce individual discretion and increase the likelihood that a task will be completed efficiently.

Therefore, granular certainty more fully interrogates the ideas that brought cabinet logic into the office. It signals how the American business imagination embraced turn-of-the-twentieth-century ideas of efficiency and a modern conception of information. Efficiency, with its focus on planning and productivity, increased the need for businesses to have information at hand; granular certainty emphasizes the historically specific conception of information required. Practices critical to the aspirations for the efficient handling of information circulated under the label of *system*. Loosely defined, system was intended to bring procedure and precision to the recording, storage, and circulation of information.

The chapters that make up the second part of this book explore how the vertical filing cabinet is imagined and presented as part of the system that made loose paper work as paperwork. The information needs of corporate capitalism made paperwork an intervention to facilitate efficiency and control. This chapter examines scientific office management's conceptualization of system and the attempt to deploy it in offices via filing cabinets; subsequent chapters explore how these ideas were simplified in promotional materials and how system was enacted through a disciplined female body that, as a female clerk, was subsumed into a collective corporate logic. Therefore, the following chapters further elaborate the argument that storage is not a neutral practice but is instead shaped by particular technologies, practices, and ideas.

Systematic Management

Some knowledge of the emergence of management as a profession in the last quarter of the nineteenth century, a period when "manager" developed as a job performed by someone who was not the owner of the business, is crucial to an understanding of how granular certainty made the filing cabinet both inevitable and necessary. Within the U.S. business world (and quickly exported beyond it), management made system and efficiency (and the logic of granular certainty) critical to the organization of industry and commerce. Contemporaries used management and efficiency to claim the late nineteenth-century period of capitalism as modern and unique, but, as critical as management and efficiency were, they were not exclusive to "managerial capitalism."[6] As Caitlin Rosenthal argues, capitalism at the scale of the railroad and telegraph was not required for the development of new management techniques. On slave plantations, the need for profit, separate from claims to efficiency, created hierarchies similar to the multidivisional form of the corporation. A standardization of accounts that enabled a form of separation of ownership and management and a commitment to the productivity analysis, later attributed to scientific management, could also be found on slave plantations.[7] Techniques identified as constitutive of managerial capitalism also existed prior to the appearance of managerial capitalism in state practices, in small-scale commercial activities, and in certain homes where from the mid-nineteenth-century domestic advice books advocated ideas of system.[8]

This prehistory of techniques is not intended to minimize the impact the new profession of management had on the business imagination during the last decades of the nineteenth century. The writings of early managers produced a set of ideas that Joseph Litterer names "systematic management"; he asserts that scientific management was an offshoot of these ideas.[9] This literature focused on perceived disorder in the organization of firms, especially those increasing in size. Not content with describing problems, the authors also offered solutions. They argued that introducing "systems" established control through integration and internal order, invoking a machine logic to explain system and control. This is not surprising given that many early managers were trained engineers whose technical responsibilities extended to new tasks that, as David Noble observes, shifted their focus from "the engineering of things to the engineering of people."[10] This movement between engineering and management continued into the first three decades of the twentieth century, when between two-thirds and three-quarters of engineering school graduates became managers in industry.[11]

The ideas central to systematic management signaled the technocratic ambition of the new profession. Management represents a specific worldview; it is not neutral.[12] As managers, engineers brought a set of ideological assumptions that defined the modern form of the office. The "machine," with its ideas of harmony and order, was central; machine logic allowed managers to conceive of organizations as sets of problems that only managers could solve.

All problems became technical problems; all participants in an organization were designated as "rational constituents of the same system."[13] Therefore, management came to view disruption of any kind as a problem of uncertainty; technical problems and labor problems were approached as "machine uncertainty."[14]

This understanding was often presented as an issue of efficiency as much as a concern with system. Efficiency, as achieved in the economy through control, focused on making the specific steps of production visible, reducing the breadth of individual tasks in an attempt to limit the possibility of error. The faithful believed that by making specific actions visible (granular) and using that to constrain what each worker could do, it was possible to increase the certainty that a task would be completed as needed. Judgment and accuracy were prized to the extent they were used to promote the value of understanding an object in terms of its components. Breaking something down into small parts allowed it to be controlled by enabling it to be more easily apprehended, understood, and connected to something else.

Information was subjected to this logic because it was viewed as critical to management's ability to control production, distribution, and eventually consumption. The information that was critical to control emerged out of standardized practices; or, put another way, the handling of paper was subject to procedures intended to produce a standardized unit increasingly called information. As one writer explained it, control was dependent on "classification, tabulation, and comparison," hence the importance of the index card to *System*.[15] Adhering to the logic of granular certainty, the breaking up of knowledge into discrete units through classification and tabulation produced information as something that could be easily accessed, seen, and connected to other pieces of information. This was necessary because, as a New York University professor of commerce wrote in 1925, "the tremendous dependence upon records as means of securing information and exercising control, which growth in the size of business necessitates, can hardly be overemphasized."[16]

However, business literature went a long way toward overemphasizing the take-up of control and the methodical recording of information. The key culprit was a set of ideas known as "scientific office management" in homage to the influence of Frederick Taylor's

Shaw-Walker Subdivisions for Filing Letters Alphabetically

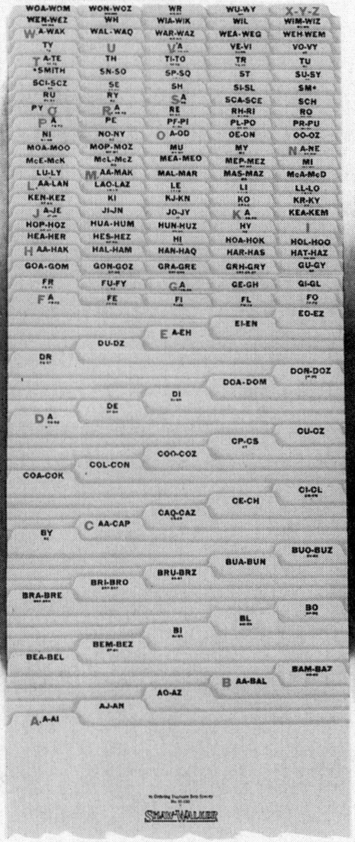

150 Subdivision Shaw-Walker Alphabet

The proper sized alphabet for three drawers of a letter file, giving 50 guides to a drawer. Back of each guide a separation is made by folders, one folder for each firm—as with all indexes. In the ordinary run of correspondence, this alphabet will subdivide about 18,000 letters properly for ready filing and quick finding.

Number 150, after one of four styles shown on pages 8 and 9, as 81-150 is the alphabet in Manila Style.

200 Subdivision Shaw-Walker Alphabet

Suitable for 20,000 to 25,000 letters, the capacity of a 4-drawer letter file. Gives 50 guides to each drawer. Smaller subdivision is false economy.
Mistakes multiply. We cannot urge too strongly the use of not less than 40 guides, and from that up to 50 or 60 guides to each drawer, depending upon whether there is much correspondence from one individual or a little correspondence from many individuals.

Number 200, after one of four styles shown on pages 8 and 9, as 81-200 is the alphabet in Manila Style.

For sizes, styles, and prices see Price List

Figure 4.2. This Shaw-Walker filing supplies catalog illustrates the guide cards and tabbed folders used to create alphabetical indexes with hundreds of subdivisions. These indexes adhered to the logic of granular certainty—that is, that breaking something down into smaller parts made it easier to manage. Shaw-Walker, *How to File Letters and Cards* (1920), Trade Catalog Collection, Hagley Museum and Library, Wilmington, Delaware.

scientific management.[17] We have already encountered its main proponent, William Henry Leffingwell. From 1917 until 1934, when he died at the age of fifty-eight, Leffingwell wrote four books, one of them a textbook that went through multiple editions, and dozens of articles and talks. He believed a businessman needed to follow Taylor and his disciples because the office had taken on an increased role in business since the time of Taylor's research and writing. Leffingwell argued that a well-organized office needed standardization to achieve control.

The early twentieth-century business world's faith in standardization affected the development of information and communication processes in the office. However, as noted in chapter 3, in American businesses this standardization did not extend to the establishment of a uniform system of paper sizes. In the United States paper did not rigorously "define the measure of its surroundings" as it did in Germany, where the A series of paper sizes was adopted in 1922.[18] As Anna-Maria Meister notes, the A4 format became a norm that "everyone could touch, everyone would be affected by, and everyone would work on"; it became a critical way in which efficiency and order circulated.[19] Although in the absence of universally accepted standardized paper formats American filing cabinets varied by two to three inches in height and depth, the recording of information on paper and its storage in folders and drawers circulated ideas of efficiency as effectively in the United States as the A4 format did in Europe.

Despite differences in paper sizes, the surface of paper was increasingly organized to standardize the recording and circulation of information. JoAnne Yates argues that the memo and preprinted forms emerged with the promise that the impersonality of standardized formats would foster efficiency and enable control in large organizations. The memo eliminated the convention of wordy formalities to create a new genre of streamlined writing distinct from letters.[20] Graphs and tables also appeared more frequently in the office in response to demands for information "at a glance"; while such techniques predate this period, they took on a more central role in the office as the speedy delivery of information gained importance during the last decades of the nineteenth century. Graphs were celebrated as "the most effective and quickest way of getting a grasp on details."[21] Similarly, tables saved time, both in their creation and in their reception. Tables meant that statistics no longer had to be embedded in prose. By presenting numbers in particular places, tables made it easier for people to find the specific information they needed, and if those tables were on cards, they facilitated the easy movement of information.[22]

Yates presents these developments as a move away from descriptive records to analytic records, which occurred when written communication replaced oral communication in the office. She argues that written communication, "the mechanism through which the operations of an organization are coordinated to achieve desired results," became synonymous with control.[23] In the companies she studied, systems of written communication developed in response to growth in organizational scale. Analytic records, memos, and reports existed to relay communication between departments, both vertically and horizontally. A "documentary impulse" encouraged managers to "gain control over their businesses by creating systems for every aspect of their processes and products—and to implement and monitor these systems via flows of written communication and information."[24]

Yates suggests that a written record established consistency across time, erasing both the need for personal memory and the authority embedded in an individual who possessed knowledge and the capacity to remember. The problem of communication, an aspect of what I have called flow, created the need for coordination. System would achieve this coordination if it prioritized "forward movement" via "sharply defined channels through which work can flow."[25] This definition of system as a track was usually paired with the idea of system as an organizational focus on people, although at times these ideas could be viewed as oppositional.[26]

Usage Scenarios

The coordination among various systems involved a shift in focus from the individual to the system, a shift that helped to provide the conceptual and physical space for machines to enter the office even if the ideas of scientific office management proponents were rarely implemented in any rigorous way.[27] Jon Agar explains the logic at play in the idea of system advocated in early management writing: "Just as a machine was made more efficient by the degree to which spare parts could be standardized and ordered off the peg, so too could a largely human organization be made more efficient if the system—explicit rules, communications, and functions—rather than the individual was considered first."[28]

Leffingwell repeatedly argued that simplifying a process would cut costs by shortening the time needed to complete it; to simplify a process was to standardize it through control. Attempts to coax organizations to act in a particular way, to adopt system in the name of efficiency, usually involved a range of technologies deployed to facilitate the movement of paper in and out of cabinets and the consequent flow of information around the office. This created a relationship between workers and information defined by speed.

The integrity of the filing cabinet simplified storage by providing a standardized or centralized place for it. The standardization of filing systems began with cabinets, drawers, and folders designed to store different sizes of paper and cards. However, the "system" that office equipment companies were trying to sell incorporated other technologies that enhanced the granularity of modern storage practices, including index systems, charge systems, cross-reference systems, and systems for transferring old documents into storage, usually at the end of the financial year (what is now called records retention). These additions furthered the process of naturalizing a specific relationship to information, an instrumental relationship to a discrete object. The aim was to create information sized to the smallest functional detail, but in such a way that managers could easily coordinate bits of information and not lose sight of the big picture that was the business or corporation.

The articulation of technology and organization is evident in the "usage scenarios" that appeared in office management books, business magazines, specialist industry publications, conference talks, filing textbooks, and office equipment company magazines. Delphine Gardey introduces this concept to describe early twentieth-century office management literature that presented the use of new technologies and an associated mode of organization as the pinnacle of progress. As she notes (similar to Agar), such scenarios fed the belief that organization and mechanization could no longer be separated, along with the idea that machines exercised power over people and places.[29]

Figure 4.3. This Art Metal advertisement shows the dominant ideas of efficiency that informed the marketing of filing cabinets. These ideas were most notable in advertisements that made filing a task defined by speed and the need to save time. Courtesy of the Fenton Historical Society, Jamestown, New York.

Usage scenarios often adopted a "before and after" structure to emphasize how various problems could be solved by an efficient filing system. This was the case with a lengthy address that Edna Cotta, the Brooklyn Edison Company's file executive, gave at the American Management Association's annual convention in 1935. Cotta offered a contrast between the company's 1920s and 1930s file systems to argue that centralization and classification created efficiency. In the early 1920s, Brooklyn Edison, contrary to the general aims of the company, did not shed light on its correspondence and records. It had no centralized filing department; rather, individual departments maintained their own files stored in vertical filing cabinets that Cotta described as "resting places for records and papers of every description."[30] These resting places housed papers not considered critical to the daily routine of the office; papers of value were stored on or in desks. The filing cabinets also contained a range of personal possessions usually found in other kinds of short-term and long-term storage units, such as wardrobes, lockers, and wastepaper baskets. With bowling balls and waders purportedly taking up space in these cabinets, some papers, perhaps unsurprisingly, ended up "resting" on bookcases or in office supply cabinets.

By the end of the 1920s the company had reorganized its files, "bringing into systematic connection all the unrelated scattered files in the company and fabricating them into an organized entity—the centralized file bureau."[31] The bureau had four hundred filing cabinets and an overflow room with sixty-four additional cabinets; departments still kept an unknown number of filing cabinets. A staff of eighty-seven women handled the company's filing needs. The daily work of the bureau involved filing 3,500–4,000 items and responding to about a hundred requests for information.[32]

Brooklyn Edison's 1930s reorganization provides an example of the wider role a filing department could play in an office. File clerks began work at 8:00 a.m. sorting incoming mail, with the goal of getting it to departments by 9:00 a.m. While large firms often had mailrooms, it was not uncommon for filing departments to sort mail. At Brooklyn Edison, this task involved tagging each piece of correspondence using an alphanumeric system that used sequential numbers preceded by letters that designated the months in which the items were received (A for January, B for February, and so on). In an outdated practice, clerks recorded letters in a register. Organizational protocol allowed departments to keep letters for two weeks before they had to return them to the file bureau. When the register revealed outstanding letters, clerks contacted the departments to request their return. If October 1937 was a representative month, the system worked. In that month file clerks registered 8,141 letters, all but 18 of which were returned to the file bureau within two weeks.[33]

The filing cabinet was not the only technology available in the Brooklyn Edison file bureau. Clerks used a "sort-a-graph" to prepare letters and files for filing, rather than simply a table, as would be common in centralized filing departments with less volume. The sort-a-graph (sold under the brand names Multisort and Savasort) was a belt with pockets or dividers sized to fit a range of cards and papers. Attached to a carriage that moved on ball bearings, it allowed the operator to use one hand to bring the required divider next to the sorting pile and place a document in the divider, all while determining where the next document had to go. At Brooklyn Edison clerks used a two-container, 1,000-division sort-a-graph. One side was alphabetical and the other geographical, with the clerk sitting between the two. Echoing the common trope of office equipment being an extension of a worker's mind or body, an observer remarked, "The operator has a low chair which slides within the compass of the two arms of the sortagraph."[34] This account makes no mention of the speed at which the clerks sorted papers, but other articles and promotional literature claimed a clerk could sort at least 1,500 cards per hour—a rate of 25 per minute, or one card every 2.4 seconds. An unnamed clerk from an unknown firm purportedly filed at the rate of 50 cards per minute, or 3,000 cards per hour.[35] As was the case with all such advertising and usage scenarios, the likely effect of using time-based data to blur worker and machine efficiency was to give employers more ways to make actual clerks feel inadequate.

Consistent with the genre of filing cabinet usage scenarios, the Brooklyn Edison example tells a story of efficiency. However, the company's experience in the 1920s was likely closer to that of other offices of the 1930s. At the end of that decade a "file specialist" from the office equipment company Globe-Wernicke candidly told a meeting of the New York File Association, "Few businesses are properly organized in their record keeping procedure. For every one organization where efficient control of records exists there are hundreds sadly lacking on this score."[36] In spite of their possibly large number, organizations with disorganized filing cabinets have

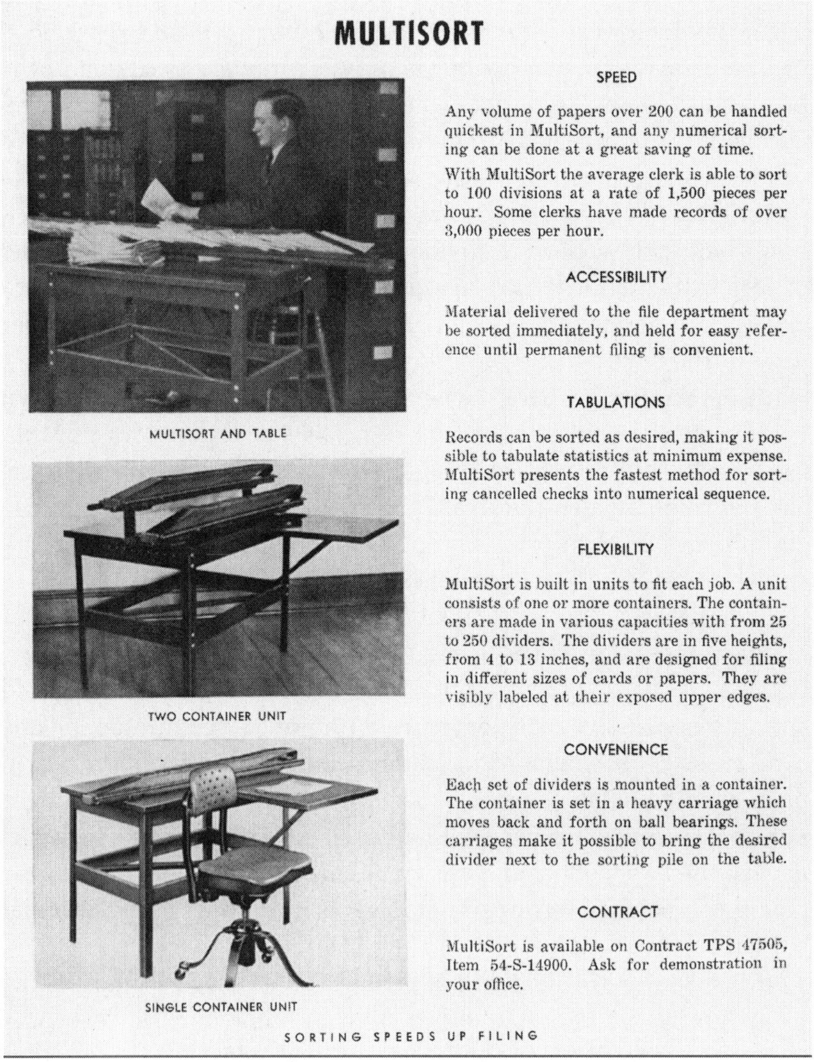

MULTISORT

SPEED

Any volume of papers over 200 can be handled quickest in MultiSort, and any numerical sorting can be done at a great saving of time.

With MultiSort the average clerk is able to sort to 100 divisions at a rate of 1,500 pieces per hour. Some clerks have made records of over 3,000 pieces per hour.

ACCESSIBILITY

Material delivered to the file department may be sorted immediately, and held for easy reference until permanent filing is convenient.

TABULATIONS

Records can be sorted as desired, making it possible to tabulate statistics at minimum expense. MultiSort presents the fastest method for sorting cancelled checks into numerical sequence.

FLEXIBILITY

MultiSort is built in units to fit each job. A unit consists of one or more containers. The containers are made in various capacities with from 25 to 250 dividers. The dividers are in five heights, from 4 to 13 inches, and are designed for filing in different sizes of cards or papers. They are visibly labeled at their exposed upper edges.

CONVENIENCE

Each set of dividers is mounted in a container. The container is set in a heavy carriage which moves back and forth on ball bearings. These carriages make it possible to bring the desired divider next to the sorting pile on the table.

CONTRACT

MultiSort is available on Contract TPS 47505, Item 54-S-14900. Ask for demonstration in your office.

MULTISORT AND TABLE

TWO CONTAINER UNIT

SINGLE CONTAINER UNIT

SORTING SPEEDS UP FILING

Figure 4.4. Remington Rand promoted its Multisort in a 1930s catalog. This piece of equipment was intended to help clerks sort papers before they filed them in cabinets. Courtesy of the Smithsonian Libraries, Washington, D.C.

left very few traces; the historical record has not captured firms' refusal to install or use efficient filing systems beyond the occasional "before" role in a tale of transformation.

Despite the fact that they represent an uncommon experience, usage scenarios are nonetheless valuable amid the limited historical records of filing systems; they illustrate how easily the filing cabinet (along with other paper storage technologies) could fit into the broader procedures and practices that defined and shaped office organization. These techniques enhanced standardization, simplification, and centralization as part of efforts to improve access to information.

The use of cabinets to store linked cards and papers highlights the obsession with particularity that instrumentalized knowledge in the name of "information" and "data." As *System* showed, precise knowledge of stock levels became a key topic in articles on ledgers, sales systems, and stock inventories published by trade journals and business magazines. The Link Belt Company, a manufacturer of conveyer belts used by mines, paper mills, and railroads, installed a punch card system and index to coordinate production and orders across its four factories. The large cards (11 by 14.5 inches) were divided into three labeled sections: "Ordered"; "Issued, Received, Balance on Hand"; and "Applied and Available." The cards were color coded by factory, and metal-tipped guides arranged them by part number or name. They were housed in a specifically designed steel record desk with six compartments. Steel vertical filing cabinets were located near the desk to store inactive cards, transferred cards, and reports of goods received.[37]

Hospitals also used different types of cabinets to manage their information needs. At Cincinnati's Jewish Hospital, the larger medical records were filed in folders in four-drawer steel letter-sized vertical filing cabinets. Each drawer contained 225 case histories, with guides separating every fifty folders. The hospital stored files that were more than three years old in the basement. Cards (including admission cards used to index the medical records) were filed in card cabinets. These were eight drawers high and were divided to accommodate sixteen files of 4-by-6-inch cards.[38] In another Ohio hospital, folders and cards were housed in the same large unit. To ensure the unit had a consistent size, the card drawers were the width of correspondence/letter drawers but divided in two parts to increase storage space.[39]

The LB file

Hospital Records

Mt. Sinai Hospital

OUR Cleveland Office has been very successful in the installation of equipment and records for hospitals, one of these installations being at the Mt. Sinai Hospital in that city.

The record room is equipped with steel vertical units. The data files, including all case histories, charts, etc., for all patients are filed numerically by admission number in correspondence size units.

The nomenclature files comprise the diagnosis and complication records, covering the entry of every patient admitted to the hospital. This record permits taking off an annual report by the classifications under this nomenclature. The above illustration shows a typical drawer in this record, which is housed in order units. The guide list is the one on which we have standardized for hospital use. They also maintain an alphabetic ledger which permits of alphabetic reference entirely independent of this nomenclature record.

The hospital also maintains an L.B. stock record—a very necessary requisite, as they carry an extensive stock of forms, supplies and equipment for hospital activities.

As a result of this installation about fifteen similar installations have been sold throughout Northern Ohio and Western Pennsylvania.

Mr. Chapman, the superintendent, is a great booster for well-kept hospital records, and has done considerable lecturing on this subject. He never fails to give reference to Library Bureau as being the proper concern to communicate with in such matters.

Boston School of Filing Alumnae Association

THE March meeting of the Boston School of Filing Alumnae Association was held March 20th in Wesleyan Hall, at 6.45 P.M.

Miss Olive Mason rendered a vocal solo accompanied at the piano by Miss Merrill of the Boston Indexing Division. Miss Merrill then gave two piano solos.

The speaker for this meeting was Miss Theresa Fitzpatrick of *The Atlantic Monthly*, who took for her subject "The Qualities that make for Success in Business."

The Hospitality Committee served refreshments; group singing and a social time were enjoyed by all.

Figure 4.5. *The LB File,* the Library Bureau's monthly company magazine, featured many "usage scenarios" involving filing cabinets. This page illustrates the sales of filing systems that combined index card cabinets and vertical filing cabinets, which were popular in hospitals and other institutions whose work generated information about individuals. Courtesy of the Herkimer County Historical Society, New York.

Hospital staff used cards to record information to be added to the comprehensive medical history files stored in vertical filing cabinets and to extract information from those files. Different information needs produced different card files, which in practice meant that a patient's record could be located in multiple ways. At the Bryn Mawr Hospital six different card files provided access to patient information: an alphabetical file of admission cards for discharged patients, an alphabetical file for the accident ward, a file to record all operations, a disease file, a diagnostic file, and a doctors' file that recorded the number of patients each physician referred to the hospital.[40] The cards could be organized according to the name of diseases or body parts. In larger collections, color provided an additional organizational structure. One hospital used buff-colored cards for patients' primary conditions, blue for associated conditions, and salmon for operative procedures.[41]

Office equipment companies went further to connect cards to the holy grail of control in their promotion of "visible indexes." These were devices that suspended index cards in individual clear pouches ("panels") attached to a metal frame so each pouch overlapped the one below to leave a strip along the bottom edge of each card visible; this strip provided enough space for a clerk to write a summary of the card's content. The visible index was celebrated as a way for a business to access, at a glance or with the touch of a finger, the information it needed to control production—although the contents were often called "data" or "facts." Boxes and books were again contrasted with a new technology and declared outdated and inefficient as ways to store information. Advertising for the visible index also took the opportunity to dismiss index card cabinets, noting the advantages of having "no inaccessible drawer files, no guides, no cards to sort."[42] However, promoted primarily as ideal for recording accounts, credits, sales, and stock, the visible index left the status of the filing cabinet unscathed.[43]

The use of filing systems to reimagine the office as a site for the systematic and efficient production of information is evident in the findings of the federal government's Commission on Economy and Efficiency (also known as the Taft Commission), which was active from 1910 until 1912. Labeled the "most comprehensive and systematic investigation that has ever been made of the national government," the work of the commission has subsequently been viewed as a foundational moment in the federal government's adoption of "records management," as it would become known in the middle of the twentieth century.[44]

The Taft Commission spent considerable time analyzing filing systems. It targeted the lack of standardized filing practices within government departments in the nation's capital and the lack of uniformity in filing systems between field offices and department headquarters. These issues were particularly apparent when commission members compared the operations of government departments with those of the large commercial enterprises they visited as part of their research, where they observed systems that prioritized the retrieval of papers in a timely fashion. This focus on storage and access shaped the representation of filing in usage scenarios that presented technology and organization as inseparable.

As a result of their observations, the commission members accepted the value of two recent developments in commercial paper storage practices: vertical filing and decimal-based filing systems. The commission recommended that government departments stop folding papers to 3½ by 8 inches for storage in document files.[45] The commission's report often referred to vertical filing as "flat filing" to distinguish it from the filing of folded documents. This label signaled one reason vertical files were believed to be efficient: they had less bulk and therefore took up less space. Saving time was more integral to the commission's second recommendation, which was that federal departments adopt a decimal filing system to organize papers by subject. With this recommendation the commission followed the precedent set by the Department of State, which had adopted such a system at the order of Secretary Elihu Root, who based his decision on the report of a private firm.[46]

The importance of a decimal filing system to efficient office work underscores the value given to indexing as a mode of search, as a system that depended on material support to make content visible in its particularity; it also offers a reminder that the origins of both the Dewey decimal system and the vertical filing cabinet can be traced to the same company.[47] The demand for specificity, the attempt to manufacture granular certainty, created a range of formats for indexes, especially alphabet-based systems. Whether for documents or cards, these indexes were sold based on the number of divisions they created in the alphabet. In the world of granular

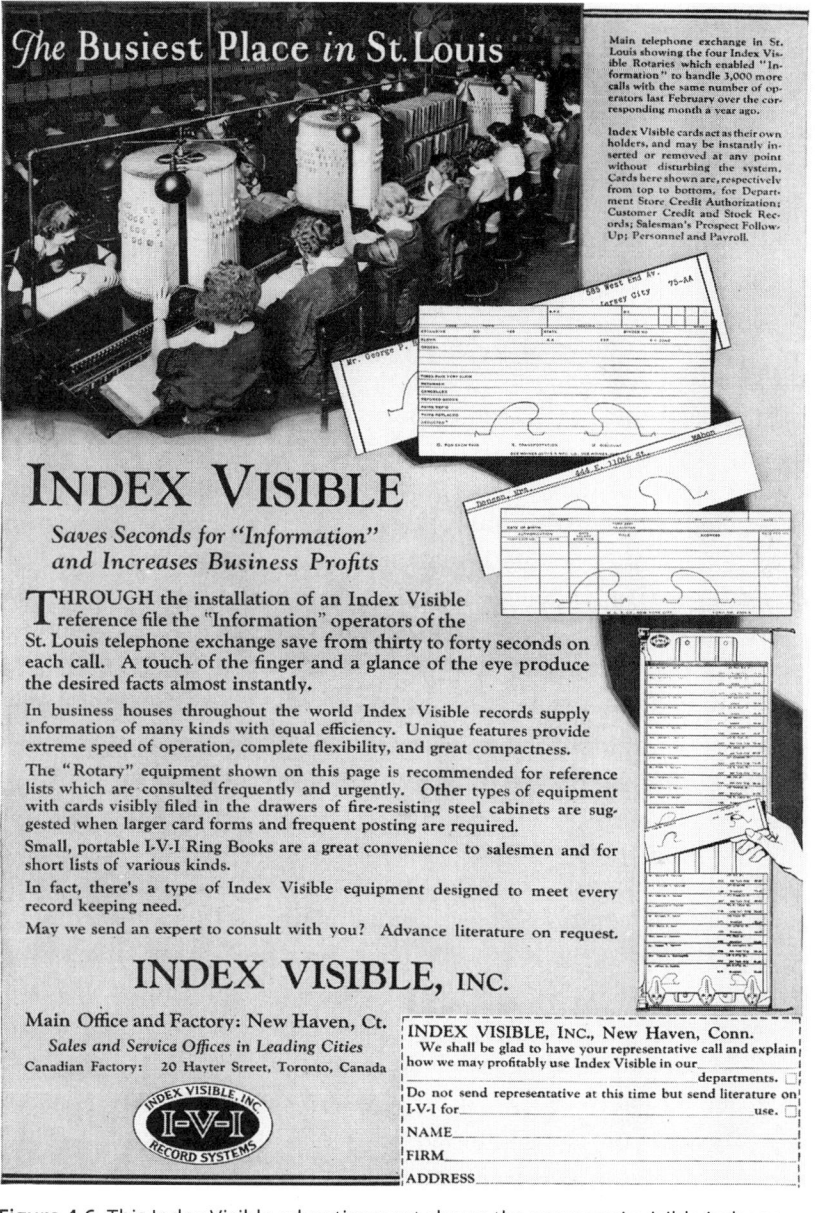

Figure 4.6. This Index Visible advertisement shows the company's visible indexes at the main St. Louis telephone exchange. These devices stored index cards so that a portion at the bottom of each card remained visible to provide easy access to frequently needed content, identified as "information." As this advertisement shows, women used visible indexes in routine labor to retrieve information. Sperry Rand Corporation, Remington Rand Division records, Subgroup III, Advertising and Sales Promotion Department (Acc. 1825), MSS 1825 III 11, Hagley Museum and Library, Wilmington, Delaware.

certainty an alphabet of twenty-six divisions was both insufficient and inefficient. An efficient alphabetical system tended to have anywhere from forty to six thousand divisions.[48]

Filing equipment companies produced a bewilderingly wide array of indexes. The Library Bureau sold its "automatic index" to a variety of business enterprises, including chemical, candy, music, paint, petroleum/oil, rubber manufacturing, and phonograph companies, as well as department stores, insurance companies, and banks. The Atlanta-based American Mills Company used four automatic indexes: a 160-division index for general correspondence, a 40-division index for credit information, and two 20-division indexes for open orders and paid invoices.[49] A hardware company in Louisville replaced its binder system with two filing collections, each requiring two automatic indexes, as old files were kept with current files. The company's City Department file used two 20-division indexes for correspondence and three 60-division indexes for orders and claims. Its Foreign Department file used two 320-division indexes for correspondence and three more for orders. A 125-division index guided the factory invoices and received bills.[50] A bond house in St. Louis used a direct alphabet index: a 250-division index for a general file, a 40-division index for a municipal file, a 125-division legal-size index for a circular file, and two 60-division indexes, one for a corporation file and one for an offerings file.[51]

Smaller businesses had specific, albeit less ambitious, indexing needs. In the world of magazine publishing, *Life* stored its photographs in a system with two main category headings: "Personality" (people) and "Subject" (places and things). The former was subdivided alphabetically by names. The latter was subdivided into general subject headings such as "Art," "Aviation," "Science," and "Sports," which were then further subdivided. *Photoplay* magazine had different filing systems for clippings and photographs. Clippings were stored in two main groups: casts, filed under the names of the plays; and biographies, filed under the names of the actors. Photographs were organized into three categories: stars, stills, and special scenes (which were classified using a subject system).[52]

The storage of photographs and clippings required not only a standardized classification system but also the assurance that paper could be retrieved and that it would not be damaged during filing, storage, or retrieval. Since clippings and photographs were too small for correspondence-sized drawers, file clerks often attached them to, or placed them within, larger and sturdier pieces of paper so they could be stored vertically (on their long edge) for easy retrieval within the information logic of the cabinet. For a number of years, *Photoplay* employees pasted biographies onto letter-sized sheets of papers and filed them in correspondence drawers. Eventually, the office started to use expansion folders for clippings. In the 1930s, *Life* rejected envelopes and folders for photographs and stored them on large cross-reference cards in an alphabetical file.[53] The New York Municipal Library made "pockets" for clippings by attaching transparent tissue paper to manila cards; this allowed multiple clippings to "be thrust into the pocket in such a way as to be still legible."[54] A more common solution to the problem of filing clippings was to place them in a stout oblong manila envelope.

The attempts to give photographs and bits of paper a standardized size were responses to the granular certainty that shaped storage in offices. In her insightful analysis of the newspaper clipping as a media object, Anke te Heesen connects such attempts more broadly to modernity. She argues that "variability and mobility, along with its capacity for separation and insertion, detachment and repositioning, and reordering and recombination, make the newspaper clipping a constitutive object of modernity."[55] These flexible capacities connected the importance of paper to the conception of information during its ascendancy in the first half of the twentieth century.

Usage scenarios from libraries provide a clear expression of clippings as an instantiation of a distinct conception of information, of discrete paper conceived as an instrumental form of knowledge. Vertical files offered a storage solution for large quantities of unbound, printed paper that libraries began to collect in the early twentieth century. A 1915 book on the operation of vertical files in public libraries explained why such files had become necessary: "The old type of library must modify itself in accordance with the new needs which the evolution of knowledge and the growth of print have created."[56] The "evolution of knowledge" was what the authors unreflectively called the move to "information." This information appeared amid a "vast flood of print" that was "ephemeral as to form."[57] The "growth" or "flood" of print included pamphlets, bulletins, reports, circulars, catalogs, advertisements, and newspaper and magazine articles. In libraries, such items became a problem because they constituted "the vast mass of material, daily growing larger, which furnishes the most recent and up-to-date information unavailable in books and unindexed in magazines."[58]

The solution was for libraries to place paper not bound in books into folders and filing cabinets designed to store business correspondence. Inside these cabinets folders were arranged using an alphabetical (or "dictionary") subject system. Headings were written on folders, and cards were used to list subject headings in larger vertical files. How-to literature recommended that items be arranged in a folder with the largest items at the back to facilitate selection. If clippings had long-term value, they should be mounted on manila cardboard; other clippings identified as having short-term or "questionable" value were to be stored loose in folders or in envelopes placed in folders. Regardless of how paper was stored, advocates promoted the value of the vertical file as a way of centralizing and standardizing documents and paper. The vertical file was described as "the most useful of all of the library's information opportunities" because it gathered "a mass of miscellaneous matter which cannot be efficiently kept in any other way."[59]

Responding to the demands of businesses, office equipment companies manufactured different-sized drawers and cabinets to accommodate the range of loose papers used to record information in offices. As a result, employees in larger offices rarely had to adapt the available equipment to their needs, as workers in libraries and smaller offices often did. In the early twentieth century, banks started to replace bound books with cabinets that were sized to house different types of documents. Parallel to the adoption of loose-leaf binders, ledger books were deconstructed into ledger cards. Different-sized vertical cabinets accommodated these ledger cards.[60] Office equipment companies produced cards designed for signatures (individual, partnership, corporation, power of attorney, and so on) and the cabinets to store them in. These cabinets had drawers divided into two compartments. Not only were the cards considered easier to handle than a bound book of signatures, but they were also presented as more private, because a person searching for a signature on a card in a well-guided drawer would look only at the specific signature needed, rather than at multiple signatures on a page in a book. Checks could be stored in drawers placed on tables, but more frequently they were stored in regular-size vertical cabinets (52 inches high by 27 inches wide) that held eight or nine drawers rather than the four or five that would be in a vertical file. According to one manufacturer, an eight-drawer cabinet offered "225 filing inches in a little over 2 square feet."[61] These drawers could also be incorporated into demi- or horizontal units along with correspondence-sized drawers.[62]

The types of unbound paper common to architecture and engineering firms were harder to fit into the size constraints of a rectilinear cabinet. That did not stop some offices from using regular-sized vertical filing cabinets to store folded blueprints, however, even though the act of folding had been rejected as inefficient in other types of offices. Advocates of folding focused on blueprints, which were made of thick paper that could be folded without much risk of damage, unlike tracings or drawings (although some outliers did suggest folding tracings). Various techniques for folding blueprints were recommended. For example, an architect could use a glass paperweight to crease the folds, in order to avoid paper cuts.[63] A set of instructions published in *Architectural Record* outlined a six-fold method. The author claimed this method could be used for drawings of any size, but he generally recommended vertical filing for blueprints measuring 12 inches by 16 inches; for larger drawings, he suggested flat filing.[64] Once folded, blueprints could be filed in a cabinet along with any correspondence associated with the job, so that all related materials were gathered in one place.[65]

Most firms, however, stored unfolded blueprints in specially designed cabinets. By the early twentieth century, the majority of engineering and architecture offices had replaced their pigeonhole files and racks (similar to those that libraries use to drape newspapers over) with these cabinets. Blueprints were often stored flat in 2-inch-high drawers, usually in five- or ten-drawer cabinets. A ten-drawer unit could hold about two thousand blueprints. Firms could partition these drawers to accommodate different sizes of drawings—a 40-by-28-inch drawer could be partitioned to create up to eight sections.[66] However, flat storage could damage unbound paper, especially when a person tried to remove a drawing from the bottom of a stack within a drawer. The manufacturers of flat files sought to counter this problem with various devices designed to "lift" drawings to allow easier access to those beneath them.[67]

Cabinets that could suspend or vertically store tracings and blueprints offered additional solutions to the drawbacks of flat filing. By the 1920s, some office equipment companies produced cabinets with frames that could hang blueprints, tracings, and drawings. Paper was placed in large envelopes or pockets arranged alphabetically or numerically and suspended from one edge in racks. So-called plan files used a compressor to keep files upright, which meant they did not have to be fastened at the top; the Packard Motor Car Company in Detroit used Art Metal's plan files to store more than 500,000 drawings.[68] The Mammoth Vertical File manufactured by Yawman and Erbe featured a temporary table that operated in the style of a gateleg table and appeared when the cabinet was opened.[69] Globe-Wernicke's Cello-Clip Map and Plan File offered a unique solution that involved suspending drawings from rods with special patented clips. Each rod could hold 750 sheets, and the "cabinet" that held the rods resembled a wardrobe.[70]

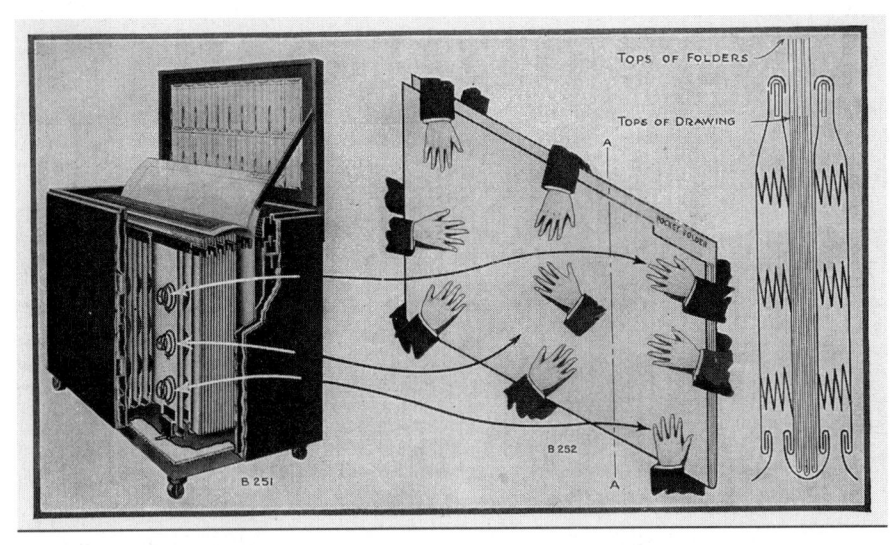

Figure 4.7. This 1928 Art Metal brochure promoted a filing cabinet designed to store blueprints and drawings. As with all filing cabinets of the period, compression remained a problem. In this image Art Metal used hands to represent the springs that compressed blueprints in the cabinet. Art Metal, *Planfiling* (1928), Lake Mohonk Mountain House data file (Acc. 2280), MSS 2280 10 01, Hagley Museum and Library, Wilmington, Delaware.

Mammoth Vertical File

A File and Drawing Board for Blueprints, Drawings, Etc.

THE "Y and E" Mammoth Vertical Wood File has a capacity of from 700 to 1,000 big blueprints, drawings, tracings, maps, charts, patterns or other large sheets, still it takes only about four square feet of floor space when closed.

Each file comes complete with 20 file pockets and 8 index cards. Extra file pockets, label holders for file pockets, casters for base legs, and metal binders for files can be supplied if desired.

No. 430. Mammoth Vertical File. (open)

No. 430. Mammoth Vertical File. (closed)

The papers are filed flat in non-actinic "folders" and indexed as illustrated below. Thus any drawing can be found instantly—unfolded, unrolled, uncreased, untorn, clean and flat.

The file opens much like an ordinary drawer-file, but the front also lifts up by patented construction to form a substantial table top where the drawings can be conveniently spread out for reference.

In this position two small brass posts rise up automatically, preventing papers from sliding off the table.

The folders, of course, are of special mammoth size—to hold drawings up to 44x30", or 48x36" according to the size of the file ordered. They are constructed with springs so that small and thin tracings do not sag at the bottom of the file.

Simple and efficient index on under side of cover makes finding of prints a simple matter.

Style No.		For Drawings	Outside Dimensions Wide	High	Deep	Shipping Weight
430		44 x 30" or smaller	$49\frac{5}{8}$"	$40\frac{3}{4}$"	$14\frac{1}{2}$"	294
430	Folders	Additional folders				
836		48 x 36" or smaller	$53\frac{1}{16}$"	47"	$14\frac{1}{2}$"	350
836	Folders	Additional folders				

Figure 4.8. This page from a 1923 Yawman and Erbe catalog marketed the company's Mammoth Vertical File for the storage of blueprints. When opened, this cabinet formed a table on which the user could consult the selected blueprint. Yawman and Erbe, *Wood Filing Equipment* (1923), Trade Catalog Collection, Hagley Museum and Library, Wilmington, Delaware.

Overlapping concerns about information and the care of paper extended beyond commercial offices. In these offices the filing cabinet remained the main vehicle through which a faith in granular certainty, the belief that information had and needed a proper place, became common sense. Usage scenarios continued to celebrate efficiency and naturalize the blurring of machine and organization.

By the late 1920s, a vertical filing cabinet had become a common sight in school offices. A survey of 522 high schools found that the filing cabinet was the second-most-used "labor-saving device" in a school office, behind the typewriter and (to the surprise of the survey authors) just ahead of the telephone.[71] Identifying the filing cabinet as a way to improve "personal efficiency," the authors viewed it as important to the "necessary" establishment of system and organization. They explicitly invoked the practices of commercial offices as they repeated oft-stated ideas from office management literature: "Organization and system are facilitated by the use of mechanical devices. Time is economized; the worry caused by details is relieved; facts are made available; 'hunches' and guesses are obviated. The mechanical devices aid the administrator in applying scientific methods to his problems."[72]

Filing cabinets also appeared in classrooms, and they were even offered as integral to a new (and likely obscure) method of teaching contemporary poetry. This approach focused on the subject matter of poems, not on authors or genres. The filing cabinet served as a place to store poems gathered from a range of sources, organized by subject. Promoting efficiency as an important goal of reading, advocates for this way of teaching poetry believed that it encouraged students to read rapidly for pleasure rather than to study poems intensively for long periods of time.[73]

Filing cabinets were also adopted in the offices of individuals who did not need to store paper to coordinate information across larger organizations. Many users celebrated these filing cabinets as producing the same efficiency that existed in large commercial offices. An article in an education journal presented a detailed description of the author's personal study as a system centered on office furniture and equipment. The author, a Smith College professor, discussed desks, chairs, and shelves before noting, "Perhaps the most useful part of the system is the filing cabinet." Within the cabinet, he designated separate drawers for lectures and correspondence. The latter he divided into active and inactive correspondence, which he further subdivided into various categories: personal, publishers, committees, departmental, student. The description suggested that he used a horizontal filing cabinet, as it mentioned smaller drawers for a daily memoranda file, a student follow-up system, and a card catalog.[74]

A clergyman's filing cabinet shared a similar structure. Letter-sized drawers contained sermons and supporting material (such as clippings and notes from books). Guides divided the material into categories and subcategories. Theology, for example, was subdivided into dogmatic, ascetic, and speculative. In a horizontal unit marketed to clergy in Boston, smaller drawers accommodated different sizes of cards: 3-by-5-inch drawers for notes referring to books and their location in the clergyman's library, 4-by-6-inch drawers for index cards for parish members, and 8-by-5-inch drawers for parish pledge cards.[75]

According to James McCord, the author of an early textbook on filing, the increased use of filing systems meant that "various improved devices were put on the market for the segregation and housing of various records not bound, finally resulting in the universal adoption of the card cabinet and vertical file."[76] To segregate unbound paper was to maintain its discreteness, to emphasize its particularity. Accommodating paper to the information needs of the office, partitions in cabinet drawers enhanced the conception of information as a discrete object—an object that was easily grasped.

Inside file drawers, paper was isolated and given a "proper place" that stabilized information in its particularity.[77] The partition enabled retrieval through visible gaps. This was a rearticulation of the logic of the archive: "an aesthetics of perception, a discriminating gaze, through which an event can be isolated out of the mass of detail and accorded significance."[78] The rearticulation was the archive being captured by the needs of corporate capitalism, as the "discriminating gaze" of the archivist became the granular certainty of the cabinet, a disembodied gaze that isolated information. By segregating papers, the filing cabinet provided epistemological support to information as a specific unit. This articulation of the archive and efficiency often took the form of claims that a well-organized cabinet created information that was "alive," in contrast to mere storage, which was characterized as "dead" or at least "dormant." In the 1930s, celebrating the use of a decimal file system, an advertising agency president noted that "information does not rot in our files, it is continuously shuttling back and forth between file and every day business use." Claiming that "within a few minutes" a clerk could retrieve sales presentations from thirty different companies in thirty different fields, he concluded, "Thus our file lives."[79]

Early twentieth-century business literature promoted standards and procedure as critical to the specificity and simplicity on which granular certainty relied. Following this logic, knowledge had to be broken down into the "simplest unit," a unit that increasingly provided the definition of information. In the office, standardized forms, memos, and routines for filing broke down knowledge into information. Using cards to record and store information in a tabular format acknowledged information as a discrete unit similar to the division of file drawers. This specificity in the name of simplicity repeated the logic inherent in Frederick Taylor's analysis of labor. Taylorism sought to extract the "one best way" from the labor embodied in the work of a craftsman.

Claims to rigor and certainty taken from science were used to link granular certainty to a shift from individuals to systems. The "scientific" approach to management was intended to move business away from a mode based on chance, what Leffingwell referred to as "that non-descript, half-presentient, half guesswork, popularly known as 'hunch.'"[80] Office management literature frequently made this point by invoking a very particular take on nineteenth-century business, which in this telling was populated by sole proprietors who had "intimate contact" with the details of their businesses. In contrast, twentieth-century business, with the corporation as its poster child, needed "an organized information system that would supply information that was not only timely, but abundant and accurate as well."[81]

These assertions are a reminder that efficiency was a temporal intervention, not only because it sought to reduce the time it took to complete a task, but also because it prioritized knowing what needed to be produced and when. Planning became critical as day-to-day businesses reoriented themselves to think about the future.

The "modern businessman" worked in a world in which he was told he needed details on a daily basis to plan for a future in which his business would be efficient and productive; information became the source of legitimacy for the process of decision making.[82]

The increased certainty that information purportedly introduced to business via planning and predictions was applied to the storage and retrieval of that information. This included the work involved in the new conception of filing as an efficient form of labor. The faith in filing cabinets' ability to orient workers to modern conceptions of time and productivity depended in part on the use of mechanization to implement procedure and simplicity, as well as the discipline associated with "personal efficiency." In the introduction to his 1926 edited volume on office equipment, Leffingwell noted that "for every phase of office work which involves repetitive effort, or which is the source of delays of inaccuracies, or which abounds in lost motions, there is today some device or some type of equipment, designed to reduce the amount of labor, to increase the accuracy of the work, and to produce the desired results in a more efficient manner."[83] Advocates argued that the filing cabinet could offer this certainty because it lessened the need for workers to think, or at least it changed what it meant to work with knowledge, to know, and to remember.

Chapter 5

AUTOMATIC FILING
Memory for the Machine

The June 1918 issue of *Steel Filings,* Art Metal's company magazine, included a poem about the filing cabinet. Its opening stanza set the tone for the doggerel:

> Oh, that great and grand invention;
>   Oh, that labor-saving plan;
> Oh, that object without motion
>   Without fingers, brains or head;
> Oh, that masterly assistant to the busy business man,
>   That can tell you in a minute what the other fellow said.[1]

Writing under the pseudonym Impressions, the poet extolled the virtues of the filing cabinet in ways that had become fairly common in advertisements, catalogs, and office management books. First, the contents of a filing cabinet were identified as "information," which was given the attribute of efficiency, in this case described (later in the poem) as "always perfectly precise." Second, the filing cabinet was presented as a laborsaving device. In a working week, understood in hours and minutes and not just days, the "busy business man" did not waste time looking for documents. A filing cabinet could save time because it told him, "in a minute," what he needed to know. Third, the dependability of the filing cabinet was frequently explained in terms of a "memory" or "brain" (located in

The Filing Cabinet

Oh, that great and grand invention;
  Oh, that labor-saving plan;
Oh, that object without motion,
  without fingers, brain or head;
Oh, that masterly assistant to the busy
  business man,
That can tell you in a minute what
  the other fellow said.

When you get a letter read it, answer it and mark it plain
For the section of the files where
  it must go.
It will be there when you need it, if
  you must refer again,
For the files will tell you all you
  want to know.

In a day of eight short hours, in a
  week of six short days,
There are many, many minutes it
  conserves to other uses,
And in oracle-like fashion, always
  right the things it says,
Always perfectly precise the information it produces.

There is no colossal genius in the
  busy marts of trade
With a great gigantic memory that
  can do the things it can.
It is only a bit o' steel, yet no brain
  was ever made
That could wholly supersede it with
  the busy business man.

When you get some information first
  write it out, then file it
In the section of the files where it
  must go.
If you gather up your data and then
  carefully compile it,
When you forget the files can tell
  you what you want to know.

It's a great and grand invention, it's a
  labor-saving plan,
It's an object without motion, without fingers, brain or head,
It's a wonderful assistant to the busy
  business man,
For it tells him in an instant what
  the other fellow said.

—*from* IMPRESSIONS

Figure 5.1. The June 1918 issue of Art Metal's *Steel Filings* included this poem celebrating the filing cabinet. All major office equipment companies had at least one in-house magazine, and these publications often featured poems, jokes, and inspirational quotations. Courtesy of the Fenton Historical Society, Jamestown, New York.

the system) that was more reliable than the minds of people, who tended to forget too easily.

This chapter explores the logic that presented the filing cabinet as a laborsaving device. The poem by Impressions included a stanza acknowledging the work involved in preparing to file a letter. But that only served to emphasize the final stanza's assertion that a filing cabinet could do its work "without fingers, brains or head; / It's a wonderful assistant to the busy business man." The filing cabinet was a wonder because it performed work "guaranteed" to provide information on "what the other fellow said," a task that fit within the dominant logic of granular certainty and the gendered hierarchy of office work, where what the man said mattered and the work of a woman to serve the man was largely downplayed.

However, when advertisements ignored the labor of filing, they attributed the filing cabinet's wisdom not to the generic "oracle" credited in this poem, but to a very specific American deity: the machine. Mediated by the metaphor of the machine, the filing cabinet could be counted on for its speed, efficiency, and order.[2] If, as dictionaries inform us, a machine is a device of interrelated parts that function together through mechanical or electrical power, a filing cabinet fits uncomfortably under the label of "machine."[3] But the filing cabinet was understood to be a machine in the business imagination of the early twentieth century. For systematic management, all business was a machine, and a machine that should run efficiently. The filing cabinet became one of many turn-of-the-century machines that promised to independently perform work previously done by people. I begin this chapter's exploration of the conception of the filing cabinet as a machine by examining the presentation of the cabinet as a form of automatic memory, both in advertising and in office practice literature. I conclude the chapter with a discussion of filing as a feminine form of the "information labor" that emerged as a consequence of the reimagination of office work through the operation of machines.

Automatic Machine

The word *automatic* came into general use in the nineteenth century, derived from the word *automaton*. Lisa Gitelman argues that the increasing use of "automatic" carried with it "lingering connotations of resolving the organic and the mechanical—of human forms and functions built into machinery and of mechanical responses by human beings."[4] Office equipment manufacturers sought to convey those connotations when they deployed "automatic" to sell filing cabinets, and when they anthropomorphized the cabinets and their contents. One writer claimed, "In a large file of about 150 four-drawer units, the conspicuous red, blue, green, and white labels fairly cry out their divisions."[5] Another simply stated that an effective filing system "will explain itself automatically."[6]

In "crying out" to a person or "explaining" the location of a letter, a filing cabinet became automatic because it blurred human and machine by taking on the mental work of remembering where papers were. By anthropomorphizing filing cabinets, advertising underscored the move away from the individual to a system that occurred in the name of efficiency. A filing cabinet could "remember" because it replaced personal memory with objective procedures, with a system.

The move to system brings to the foreground the cabinet logic inherent in the structure of the filing cabinet. The mundane reality of the filing cabinet as an automatic machine depended on partitions, which took the form of paper tabs, folders, and guide cards. Therefore, to label the filing cabinet automatic brought "intellectual furnishings" into the project of systematic management and the celebration of efficiency.[7] In this version of intellectual furnishings, the reimagining of furniture as a machine made explicit the belief that furniture was more reliable than people for effective memory storage and retrieval.

To convey the precision, consistency, speed, and reliability that they believed would make filing seem sufficiently automatic, office equipment companies introduced complex systems of tabs, guide cards, and manila folders to create multiple subdivisions, which became the basis of claims to automatic filing. The indexes offered combined alphanumeric classification systems. Some used alphabetical guides to find folders and numerical guides to file folders; others used numbers to subdivide alphabetical categories. Tabs staggered across the tops of guides made the alphabetical and numerical systems visible. Different colored tabs divided the horizontal plane of the file drawer into three sections: an alphabetical guide for customer folders, miscellaneous folders, and folders for high-volume correspondents.

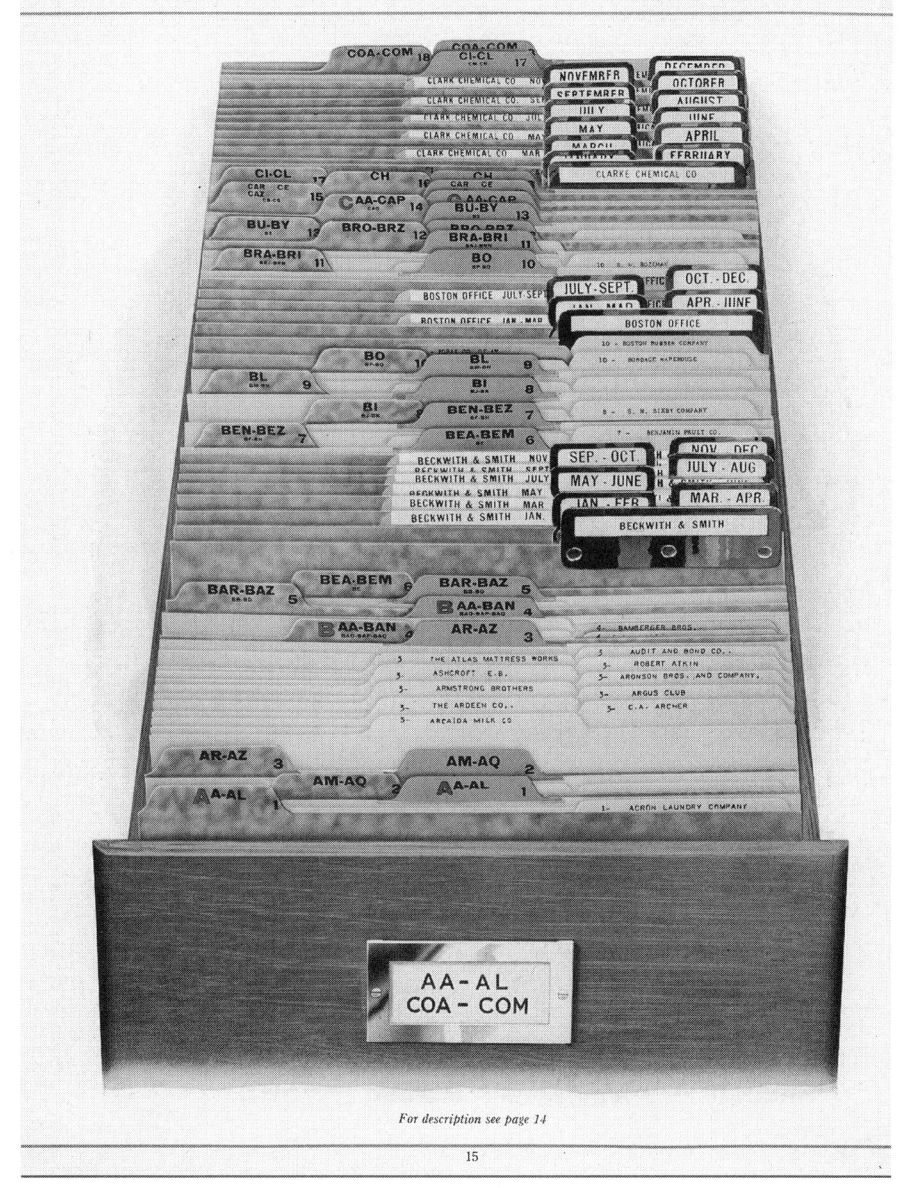

Figure 5.2. This catalog page shows Shaw-Walker's Super-Ideal Index, one of the many alphanumeric systems that filing equipment companies produced in which different colored tabs were positioned to help clerks identify where to find documents. The increased certainty attributed to these indexes allowed companies to present filing cabinets as machines. Shaw-Walker, *How to File Letters and Cards* (1920), Trade Catalog Collection, Hagley Museum and Library, Wilmington, Delaware.

Office equipment companies introduced these paper technologies as part of their attempts to confirm the filing cabinet's status as a machine and solidify its reliability relative to people. Shaw-Walker called its filing system the Super-Ideal Index and described it as "the old, simple alphabetical method made mechanically perfect."[8] Other companies also used names for their systems that sought to convey reliability and accuracy, which they explicitly connected to the imaginary of a machine through their advertising copy; examples include Globe-Wernicke's Safe-Guard and Yawman and Erbe's Direct Name. The Library Bureau justified labeling its version "automatic" by noting that it did not require a separate index or registry; everything needed to find the proper place for papers was inside the file drawer. It promoted its LB Automatic Index as doing the "two things one wants in a filing system—the first is accuracy, the second is rapidity (speed efficiency)."[9] One way the index did this was through its numerical control, which the advertisements presented as a "self-checking" device. When salesmen discovered that at first sight some potential customers found the LB Automatic Index too complicated to bother with, the company responded by outlining its merits as simply as possible through the following scenario, which appeared in promotional booklets and catalogs: "The file clerk about to put the correspondence of Jones marked 113 carelessly into the files, sees that her hands rest on the folder marked 117. And, naturally, she is checked at once, and is reminded that Jones's correspondence should go into a folder marked 113. The L.B. Automatic actually makes it difficult to perform an error in filing."[10]

To give the filing cabinet the ability to "remind" clerks that papers had been removed, office equipment companies introduced "charge," "out," or "substitution" cards to take the place of papers when clerks removed them from a filing drawer. The cards, larger in size and/or different in color and thickness from the folders, were printed with spaces where clerks could to briefly note the content of the absent papers, who requested them, and the date they left the filing cabinet.[11] Outside the cabinet, a route slip could be attached to a file to "prevent a paper wandering aimlessly around the office in the hope that it will eventually find someone to answer it."[12] Paired with the charge cards in the filing cabinet, route slips extended the scope of the cabinet and maintained links between files and the centralized filing system.

As a machine, a filing cabinet could make a worker more machine-like. The author of an early book on filing applied the idea of functioning automatically to clerical workers, defining it "as a matter of habit to perform at the proper time every individual duty that is necessary to keep the system always in complete and efficient working order."[13] Although that author suggested that a trained supervisor could model a relationship between worker and system, most subsequent writers of how-to-file instructions echoed the faith in equipment that manufacturers sought to convey in their advertisements. Filing equipment was understood to "provide every sort of safeguard to prevent mis-filed papers."[14]

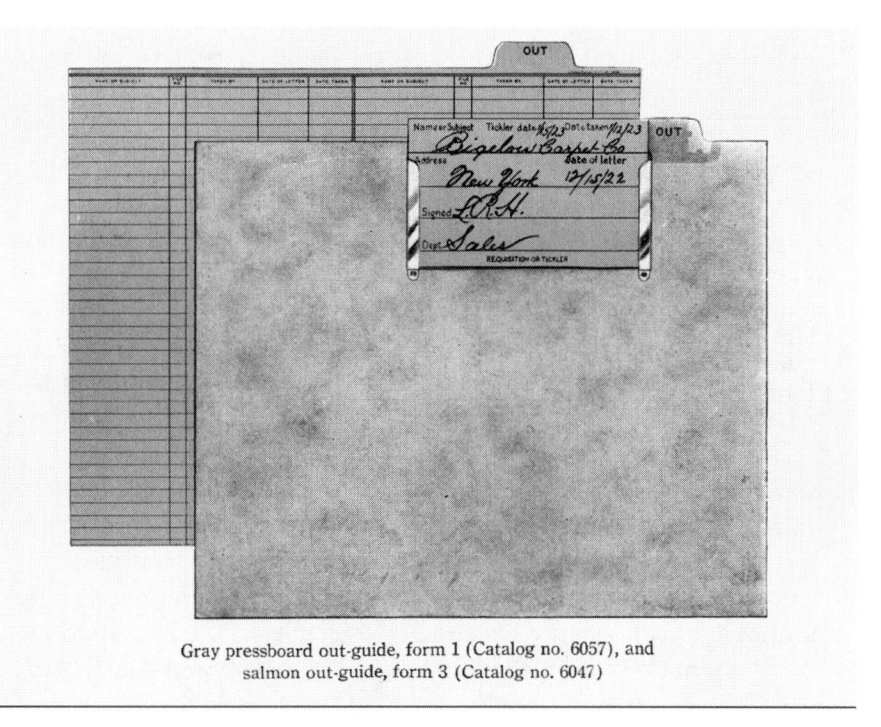

Gray pressboard out-guide, form 1 (Catalog no. 6057), and salmon out-guide, form 3 (Catalog no. 6047)

Figure 5.3. This image from a Library Bureau catalog illustrated a section on how to use charge cards. These were introduced to manage the movement of paper out of filing cabinets in response to the anxiety generated by the shift from keeping records in bound volumes to storing loose paper in filing cabinets. Author collection.

Automatic Memory

The filing cabinet, imagined as a machine, provided the large capacity for memory ("gigantic" according to our poet) necessary for the modern business world, where efficiency depended on planning that required a constant flow of information. The conception of the filing cabinet as possessing a memory made it possible for a business or office to have a memory, something office management authors regularly noted, along with the assumption that this was a perfect memory. A professor of accounting considered it necessary to note such a claim was figurative before he named a chapter "The Office Memory" and identified files as "another means of keeping alive the office memory."[15] Viewing an office as having a memory reduced memory to the act of recall; papers stored in a filing cabinet were there to be used.

In her 1923 book *Filing Department Operation and Control*, filing instructor (and former clerk) Ethel Scholfield argued that "memory is a factor no longer to be depended upon," especially given the demands of accuracy, speed, and flexibility confronting modern businesses. With the increased scale and expectations of twentieth-century corporate capitalism, an individual businessman could no longer depend on his memory. Scholfield noted that replacing this memory "presupposes a thoroughgoing automatic system for the association of ideas." Posing the question "Can such a thing be secured by mechanical means?" she immediately answered in the affirmative: "Experience tells us 'Yes.'" Therefore, Scholfield called vertical files an "automatic memory."[16]

Scholfield presented the problem of "association" as the need to improve on human memory so it could work at the scale of a modern business; such improvement would come from outsourcing memory. Despite her metaphorical reference to memory, Scholfield's response centered on procedures, not memory; the assumption that a functional equivalence existed between the human mind and the filing cabinet or the office naturalized the system and control that were labeled "automatic memory." The solution to the problem of association was twofold. Before the act of association was mechanized, it had to be made "scientific." In this role, science named an objective system of classification, as authors such as Scholfield channeled the late nineteenth-century "enthusiasm for classification as a technology of search and retrieval."[17] It produced groupings based in agreed-upon rules that could be applied throughout a business, replacing the "ingenious subdivisions" of an individual's mind; hence, for Scholfield, *science* and *routine* were synonyms. Files, in her brief discussion of science, became the "new scientific impersonal memory of the business organization."[18]

Scholfield asserted that the routine application of classifications in modern offices required "mechanization." In her conception of the office, to be automatic was to apply objective rules on a large scale. Therefore, as I have noted, the response to the large volume of paper in an office was the introduction of classifications that subdivided the paper so information could be found easily. Information was critical to Scholfield's explanation of the logic of filing. Science, the backbone of filing, was defined as "classified information." Filing as automatic memory turned the "association of ideas" into the "coordination of information" and thus connected it to twentieth-century capitalism's investment in control. At the increased scale of the modern office, information was the objective substance to be found in a filing cabinet.

When mechanized, the association of ideas was no longer a mental connection, a recollection that linked a memory and an object. With the object being information, not ideas, it became necessary to make it coordinate, to place information in a proper position relative to other information. Adhering to the logic of the metaphorical use of memory, in replicating human memory on a scale greater than human capability allowed, this coordination would be "so automatic as to seem almost human."[19] That the memory of the filing cabinet was in fact based on procedure and system speaks to the lingering connotations of "human forms and functions built into machinery" that Gitelman argues accompanied the popular uptake of "automatic" in the nineteenth century.[20]

In this chapter, I am interested in examining what happened to "memory" when authors like Scholfield presented it as the product of a machine, when it became the solution to demands for accuracy, speed, and flexibility in the office—when memory was co-opted by efficiency. Extending the concept of memory to the filing cabinet continued a historical metaphor that represented memory as a storehouse, a preneurological understanding that memory was a faculty of storage and recall.[21] Memory operated by storing impressions somewhere in discrete spaces of the brain; for a business imagination besotted with granular certainty, this was an appealing metaphor.

An explicit example of how promotional rhetoric used granular certainty to connect filing and office equipment to the brain and memory can be seen in a catalog of "filing devices" distributed by the Chicago-based company Flexifile. Targeting small-scale businesses with its line of "Desks with Brains," Flexifile highlighted how the pedestals of the desks could house a range of file drawers to accommodate the different sizes of paper used in a small office. The company claimed, "The compartments of the 'Desk with Brains' are—like the cells of the brain—places in which to systematically file data, letters, papers and information, ready for instant reference."[22]

The "cells" of the brain invoked by Flexifile have their roots in the metaphors used to describe memory in ancient and medieval times; this history is important to establishing the continuities and discontinuities in the twentieth-century claim to "automatic memory." The Latin word *cella* provided one of the commonly used place-based metaphors for memory.[23] *Cella* meant storeroom, but it had other meanings that connected it to the use of birds and insects to represent the relationship between memory and study; the word could refer to both dovecotes and the compartments bees made for honey. An ancient version of the former usage appears in Plato's dialogue *Theaetetus,* where pigeons and pigeonholes serve as metaphors for thoughts and memory.[24]

In contrast to another common metaphor for memory, the wax tablet, the metaphor of the storeroom, or *thesaurus,* made the relationship between structures and their content, and that between internal organization and classification, central to memory. By the medieval period, other spatial metaphors for memory included the wooden chest or box *(arca),* the money pouch *(sacculus),* and the letter case or book box *(scrinium).* These were used to represent the space in which ideas were placed in someone's mind to be retrieved through associations. As Mary Carruthers observes, this privileged an understanding of memory as recollection, where the key task was not to learn something by rote but to think by tracking down memories through connections.[25]

Memory imagined as a storage structure was inseparable from learning. An educated memory, training in the rhetorical "arts of

The "Desk With Brains"

The "Desk With Brains"

The compartments of the "Desk with Brains" are—like the cells of the brain—places in which to systematically file data, letters, papers and information, ready for instant reference.

No matter what you have to file, there are cabinets, built in sections, made purposely for it, and these sections can be built up to form the pedestals of your desk.

Illustrating Group of Regular Standard Cabinette Sections and Top Before Being Stacked Together to Form a Desk.

17 Sections from which to choose. 3 Sizes of Tops. 3 Finishes, Light Golden Oak, Genuine Mahogany and Birch Mahogany. Thousands of combinations possible.

Illustrating the Desk Set Up. All Sections Locked Together Forming a Solid Substantial Desk.
FlexiFile Detail on pages 14-15.

Figure 5.4. In its catalog of filing devices, Chicago-based Flexifile presented filing equipment, specifically its flat-top desks, as a substitute for memory and thought. Flexifile added file drawers to its desks, which provided the basis for the claim that the desks had brains. Flexifile Co., *General Office Equipment* (1916), Soda House, Call Number 2280, box 9, folder 7, Hagley Museum and Library, Wilmington, Delaware.

memory," was necessary for invention, experience, and knowledge; it provided the construction material for thoughts.[26] As Carruthers puts it, if a medieval abbot wanted to authenticate the charters of his foundation, he looked for a written document in the library; he did not search through his trained and well-structured memory. The contents of the storage spaces of his memory were "riches" valued "in terms of their present usefulness, not their 'accuracy' or their certification of 'what really happened.'" Carruthers clarifies this understanding of memory through a comparison with the twentieth-century use of the "filing cabinet model" in psychology. This metaphor makes the content of memory "documents," not "riches." The filing cabinet model identifies successful memory as "finding unaltered, unculled material . . . that, like stored documents, remains unchangingly complete and accurate."[27]

A version of the filing cabinet metaphor for human memory appeared at the same time filing cabinets were beginning to arrive in offices in growing numbers, predating the psychological model but simultaneous with claims that the filing cabinet could extend a businessman's memory or create a memory for the office. Newspaper and magazine articles increasingly mentioned the filing cabinet in stories that described what were presented as remarkable feats of recall. A 1910 newspaper article praised the ability of a detective to recognize criminals from photographs he had seen decades earlier. The writer asked, "How is it that this man after looking at a photograph once can pigeonhole it in his brain as well as in a filing cabinet and recall the same face after a lapse of twenty years even though the crook has grown a beard and otherwise altered his appearance?"[28] The references to pigeonhole and filing cabinet offered an implicit answer, and also reflected the time in which the article was published. During this period, filing cabinets were still finding their way into offices. Pigeonholes as places of storage were on their way out, following their popularity in offices in the middle of the nineteenth century. As physical pigeonholes became less commonplace, however, the word took on the meaning of a fixed category into which something is classified. This also suggests why the writer chose not to use the term *photographic memory,* which came into use in the middle of the nineteenth century. With its suggestion of an innate ability to remember without overt organization and procedure, the concept of photographic memory did not fit into ideas of system and granular certainty.

The filing cabinet remembered accurately because it made visible the system it used to organize information. This was the basis of the claim that in a filing cabinet, files could be found "at a glance." To justify their descriptions of filing as "automatic," office equipment companies collapsed visibility and accessibility as they promoted the merits of storing unbound paper vertically. The drawer labels on a grid of filing cabinets and the tabs in an open drawer offered an overview filtered through an increasingly dominant articulation of efficiency and information.[29] Thus, once opened, a filing cabinet allowed a user to find papers "at a moment's notice" or "almost instantaneously." In an era of system, these claims were quickly quantified. According to a solicited letter from a satisfied customer, "almost instantaneously" translated to twenty seconds to find one letter and ninety seconds to file five folders.[30] Statements such as these, which emphasized speed, further enhanced the representation of the filing cabinet as a machine; only a machine could provide service that was automatic.

Reducing the act of filing or retrieving papers from a file drawer to a glance depended on the tabs that made the filing system visible. Tabs gave papers their "proper places," usually based on the alphabet. A person need only summon a glance because guide cards were "the intellect of the filing machine," as an office equipment company described them in a catalog.[31] The filing cabinet's intellect was relatively simple. In practice, the "system for the association of ideas" that Scholfield championed was a very limited system. However, a system that organized names alphabetically was sufficient for the needs of most offices, especially when it was contained within a system of cabinets designed to allow the collection to grow. The integrity of the filing cabinets accommodated the provisional completeness of the system (as a loose-leaf ledger did). Therefore, the user of a filing cabinet knew where to go to find a document within the alphabetical system even if the proper place for the document moved within a drawer or into a neighboring drawer as the number of papers increased.

Other systems for keeping track of unbound paper responded to a different set of aspirations regarding memory, storage, and looseness. Rather than requiring users to remember what was already known, cross-referencing systems created a secondary memory that could produce new knowledge. These drew on indexing systems developed in early modernity when memory devices such as common placing and note-taking emerged.[32] As a result of experimenting with paper in the two centuries after the emergence of print amid the demise of the rhetorical arts of memory, scholars came to value excerpts taken from books and manuscripts; initially these notes, even if collected on bits of paper or cards, were more likely to be glued into books than to be stored loose in cabinets or boxes.[33] The indexing systems developed to organize the notes gave them value as memory devices. In contrast to the filing cabinet, this was an open system. Order was constructed through a keyword search, not through a singular prearranged system such as an alphabetical name index. Cross-referencing via keywords created new knowledge by bringing notes from different places together in a box or slipcase.[34] Therefore, as Alberto Cevolini argues (channeling Niklas Luhmann), the memory structure of these devices is a network of references to which every new entry is linked; it is intended not to preserve memories (combinations) but rather to preserve memorability (combinatory *potential*).[35] To emphasize this rearranging of elements, "the property of free motion," Markus Krajewski calls a system of note cards stored in a cabinet a "paper machine."[36]

Figure 5.5. Globe-Wernicke's advertisement for a set of file folders and guide cards plays on racist tropes to sell the product as a substitute for personal memory. The advertisement appeared on a blotter to be placed on a desk. Author collection.

In contrast, the rearrangement in a filing cabinet was limited to the act of gathering papers associated with a person or a company.[37] However, as noted, the gathering of papers in tabbed manila folders did not prevent the filing cabinet from being promoted as a machine. The idea that a filing cabinet would allow a clerk to find papers with the certainty of an automatic machine dominated advertising. In the 1920s and 1930s, Globe-Wernicke marketed its Tri-Guard files through a series of advertisements that claimed no personal attributes could trump Tri-Guard when it came to "faster filing and finding." These advertisements used racist and sexist tropes to show the inadequacy of people relative to Tri-Guard files. In one advertisement, a male boss introduced a new file clerk—an Orientalist caricature of a fortune-teller holding a crystal ball. The copy identified the clerk as "the seventh son of a seventh son," further suggesting that he possessed special powers, but his powers were presented as inferior and redundant in a modern office filled with machines. The advertisement acknowledged this inferiority through a contrast between traditional and modern cultures, explicitly stating the modernization attached to the Globe-Wernicke product ("Modernize your office—it pays"). An accompanying photograph showed a woman smiling as she effortlessly retrieved a letter from an open file drawer. Another advertisement included a drawing of a young female file clerk telling her boss that he "better get new files or a memory course for me!" Memory training was a common topic in advertisements in business magazines such as *System*, though these advertisements were usually aimed at the presumed readers of the magazines, male executives. In this advertisement, Globe-Wernicke presented such training as pointless: the "strain on a gal's memory" could be only relieved by a filing cabinet that automatically guided a clerk to the needed papers.[38]

Misfiling was impossible in the ideal world of filing cabinets with automatic memories. A dominant belief in the fallibility of labor in the machine age of the early twentieth century heightened the filing cabinet's status as a machine. From this perspective workers would always be unreliable or inefficient compared to the complete automation a machine promised. The processes of standardization critical to the definition of filing as automatic could only be partly regularized, however, because the filing cabinet was not fully automatic and because an actual human being, who was "much more unpredictable than a machine in aptitudes, skills, and personality," was always involved.[39]

Information Labor

How did the conception of the filing cabinet as an automatic memory shape the cabinet's use? The erasure of labor, or its presence as failed labor, in filing literature provides the broad parameters within which the filing cabinet's ideal operation and operator may be understood. However, to answer this question fully, it is necessary to identify filing as *information labor*. Information labor is not a distinct occupational category; rather, the term refers to a type of instrumental encounter between people and information that became more common in offices in the early twentieth century. In its ideal form, this is an encounter that requires neither thought nor interpretation and does not directly produce knowledge. A product of discourses of system and efficiency, it fits into a conception of work that depends on rational and calculative procedures. Information labor is not knowledge work. It is the manual labor associated with information, not the mental work associated with knowledge. To that end, information labor is located in a hierarchy below work that is understood to rely on thought. It developed out of Frederick Taylor's campaign to use mechanization to erase autonomy and skill in manual labor, a cause that accepted capitalism's dependence on labor that is stripped of autonomy (what Karl Marx referred to as the technical division of labor within a factory).[40]

Labor is always situated in time and space. In the case of the information labor of filing, the time and space were products of the early twentieth century. Information labor is situated in the temporality of efficiency, a belief in the necessity of speed, which marked filing as instantaneous. Its space (initially) was the modern office. The more rigorous division of work and the presence of women as low-level clerical workers denoted it as new. Therefore, information labor is a feminine mode of work; more precisely, it is a gendering practice—to perform information labor is to do woman's work. This locates information labor as a specific instance within a longer history of "the gendered context of working on, with, and around paper."[41] To emphasize the historical specificity of this context—that is, the connections among gender, paper, efficiency, and information—it is crucial to clarify the object of this gendered mode of labor: information as a standardized object that can be worked on without being fully comprehended, information that is encountered via machines.

Information

A focus on information labor highlights the overlap between efficiency's embrace of standardization and a conception of information as something discrete and particular; the information and the labor of information labor were both products of a faith in granular certainty. Early twentieth-century office equipment produced a functional relationship between information and a person; the functionality gave an important and particular role to a worker as a machine operator and therefore to her hands. Constrained by technology and intended to serve or assist the work of someone else, this relationship constitutes information labor.

In turn-of-the-twentieth-century offices, as the filing cabinet mechanized remembering, so the typewriter mechanized writing. The typewriter introduced a specific material and technical relationship within the history of writing, a history that, as John Durham Peters illustrates, shows that "writing is always material and technical, however automated or naturalized it may feel to its users."[42] Compared to the users of previous writing tools, however, a typewriter user was more overtly seen to be a machine operator. With this fusing of body and machine, the continuity of mind, body, and writing was broken.[43] The result was writing made up of words created from standardized and discrete units (typed letters), in contrast to cursive writing. As Gitelman argues, typewriters "challenged the author as agent by offering a newly mechanized, newly gendered, and self-consciously 'managed' imposition between the mind and page."[44] Therefore, in the office, the process-oriented mode of work introduced by the typewriter associated writing with a machine, which, as the modern office developed, created a gendered relationship to writing not centered on creativity (this did not happen instantly—for example, in 1904 only 21 percent of typists in federal government offices were women, although women had been working in those offices since the middle of the nineteenth century).[45] The

Figure 5.6. Advertisements for office equipment regularly showed filing to be an action done by women in the service of men. The cover of Yawman and Erbe's visible index catalog shows a woman providing a man with information, allowing him to continue his telephone conversation uninterrupted. According to Yawman and Erbe, this assistance, accomplished through the use of office equipment, would help an office achieve "control." Yawman and Erbe, *Visible Index Systems: Better Business Control* (n.d.), Trade Catalog Collection, Hagley Museum and Library, Wilmington, Delaware.

product of this mechanical mode of writing was information, not knowledge. Like the contents of the filing cabinet, it was information defined as "always perfectly precise."[46] Standardization in production and form made this information precise; it was information demarcated by its particularity.

Filing as the handling of paper offers a more direct way to grasp the nature of the information that became the object of information labor. Clerks had to pick up, hold, and place paper. James McCord, who founded the New York School of Filing in 1914, explained in his 1920 textbook how to work with paper in the drawer of a filing cabinet:

> Use your hands for selecting the proper guide and filing. Should the guide be far back in the file, draw the bulk of the other guides and folders forward by grasping at the sides. Remember that the tab is the weakest part of a guide, and while you will often have to finger it directly, remove as much weight as possible from the front of it before doing so. In putting material in the folders, lift the folders up if not entirely out of the file, otherwise papers may be placed in the wrong folders or between folders.[47]

Although the filing cabinet lacked the mechanical parts that underscored scientific management's celebration of dexterity in the factory, McCord's description illustrates the importance of hands and fingers to filing in the office; filing demanded more than the mere handling of paper, it involved selecting, grasping, removing, lifting, placing, fingering. Fingers grasped and displaced objects, isolated them, and gathered them together; they manipulated and shaped objects.[48] Guided by the "intellect of the filing cabinet," the hands of a file clerk manipulated information, not in the sense of falsifying information but in the sense of handling it, of moving information within the office. Unbound paper in tabbed manila folders offered a mode through which paper gave physical form to information as a discrete unit that had presence in the world. This was information, which, in contrast to knowledge, did not need a knower.[49] To file was to generate a sense of what information felt like, that contact with information could be shaped by the senses more than by the mind. To work with information did not require thought; the need to *not know* separated information labor from knowledge work.

Labor

Gender underwrote the distinction between information labor and knowledge work in the office. It reinforced the secondary status of information labor as something women performed; men could do these tasks, but they were things a knowledge worker would do on the side without any thought. A catalog description for Yawman and Erbe's Efficiency Desk, which included file drawers, made this point through the lack of subtlety critical to the genre of the trade catalog: "Each compartment should represent a fixed place, so that the hand of the executive will reach automatically for desired records without interrupting the continuity of brain action."[50] In this scenario a man filed, but only while he thought about something else. He did not have to think about retrieving documents. He reached over to the drawer as a matter of habit, while the file drawer, as a machine, worked to locate information for him. A file drawer remembered. It allowed the male executive to keep thinking about matters deemed productive. As an executive in the gendered office hierarchy, a man was employed to think; *his* work was the priority of the office.

When the scale of business produced so much paper it had to be stored beyond an executive's immediate reach, it was placed in a filing cabinet. Filing cabinets freed the executive from information labor by creating a mode of storage that generated work for women. Filing, isolated as a distinct job, was coded as women's work because it did not directly contribute to the productivity of the office—women worked in the office to assist men, so the men could think. As Delphine Gardey reminds us, "To be a boss is not to be disturbed; it is to be served and liberated by technology (and by others' work) rather than constrained by it."[51] The division between information labor and knowledge work was not simply about machine work. A man could operate an efficiency desk, and he could also use a bookkeeping machine, even in a period when advocates for the professionalization of accounting used those machines to differentiate bookkeeping from accounting. However, as Sharon Strom notes, in advertisements for bookkeeping machines, men were depicted using the machines individually (akin to the advertisement showing a man using the file drawer in an efficiency desk). In contrast, women were shown in groups using the machines to do repetitive work; these routine actions defined the only types of tasks they did in an office.[52]

WOOD FILING EQUIPMENT

"Y and E" Efficiency Desks

Your Workbench Should Be a Desk and Filing Cabinet Combined

THE "Y and E" Efficiency Desk enables you to keep your vital current records *at your finger tips*—classified and instantly accessible. It adds the convenience of a well-built desk to the advantages of a high-quality filing cabinet—at the price of a desk alone.

Leaving your desk to refer to distant files—requesting information from subordinates—telephoning to different departments to secure facts—all these annoy, waste time.

The upper right hand drawer, shown above, is the logical card-record drawer. It permits the desk user to hold the telephone receiver to the ear with his left hand and draw out cards or other information from this drawer with the free hand—making for speed in transmitting information. The need of "holding the line" or telling an inquirer that you "will call back later" is obviated.

Features That Speed Up Your Work

Vertical Drawers

Drawers are 11$\frac{15}{16}$" wide, 11$\frac{1}{2}$" high, 27$\frac{5}{8}$" deep inside. Partitions are adjustable so that compartments can be set up for different size material.

Card Drawers

Vertical drawers, both on the right and left side of the desk, offer an ideal place in which to keep sizable current papers, telephone books and advertising literature.

In the illustration, notice that the papers are filed *across* the drawer, so that the files face the desk user.

The Card File Drawers measure 11$\frac{15}{16}$" wide, 5$\frac{5}{8}$" high, 27$\frac{5}{8}$" deep inside. They have movable partitions so that compartments can be set up for the standard 5x3, 6x4 or 8x5 card sizes. In each compartment there is a depth of

Page Forty-three

Figure 5.7. The modern office was organized around a gendered distinction between mental work and manual labor. In promoting its version of an "efficiency desk" in a filing catalog, Yawman and Erbe devalued the feminine work of filing by presenting it as something a man could do while working on something else. Yawman and Erbe, *Wood Filing Equipment* (1923), Trade Catalog Collection, Hagley Museum and Library, Wilmington, Delaware.

Advertisements often showed women demonstrating how to use filing cabinets, but the primary role of the file clerks in these advertisements was to model a heterosexual definition of female beauty for the presumed straight male readers of business magazines and trade catalogs.[53] When the functioning of the filing cabinet was an advertisement's primary focus, the woman's body disappeared, although sometimes her arms and hands remained. If the advertisement emphasized how a filing cabinet kept paper vertical and information accessible, a pair of disembodied, but gendered, arms often showed how the filing cabinet efficiently stored paper. To illustrate how guides and tabs made information visible and accessible, advertisements showed only hands extracting files. The products being sold made it necessary to show the interiors of file drawers, albeit in different degrees of close-up. However, the composition of the images made it impossible to include the bodies attached to the hands and arms that used the cabinets.[54] Therefore, the disembodied hands and arms not only pointed out how the equipment functioned but also represented the ideal relationship between labor and technology necessary for the cabinet to be labeled "automatic"; hands or arms separated from their bodies and minds suggested that the users of this office equipment did not have to think as they worked.[55]

A disembodied hand offered a powerful symbol, because a hand is often positioned as an extension of the body, the way a body makes contact with the world; this can be extended again through the use of a tool. Hands can be seen as an extension of the brain, an extension of free will. The arm and hand, by reaching out to touch and grip, involve themselves in a process of choice and deliberation. However, in filing, the process of choice and deliberation (thinking) was transferred to the predetermined pathways of tabs. The disembodied hand emphasized this by *not* showing a connection to the body and mind. The hand of the file clerk fits into a history of labor in which workers have been identified as "hired hands" or simply "hands" since the seventeenth century.[56] As Janet Zandy argues, "Truncated hands represent, metonymically, an ignored whole, a lesser human element and species."[57]

The composition of the images in these advertisements crudely illustrated how delegating memory to a machine turned the mental process of remembering into a manual act dependent on dexterity, not thought. In a series of captions accompanying drawings of

THE GENERAL — GF — FIREPROOFING CO.

The Super-Filer Way
—Four Simple, Easy Steps Instead of the Usual Eight.

The Drawer is Opened

Step No. 1 A light finger pull opens the Swing Front, unlocks the buttonless safety latch, and the drawer coasts out on balanced, ball-bearing suspensions. The Swing Front plus the mechanical action of the Throw-Back slope the material backwards and make every guide and folder visible. Contents cannot fall forward accidentally.

Contents are Parted

Step No. 2 The contents are parted in one simple, effortless motion. As the finger touches the index tab of the required folder, the Throw-Back Compressor locks automatically against the Swing Front, releasing compression and freeing both hands of the operator. Note the orderly arrangement of the contents and the instant accessibility and visibility of the records.

Letter is Dropped Into Place

Step No. 3 The letter is dropped directly into its position—no necessity for removing the folder, as in ordinary filing cabinets. The wide angle, book-like spread permits the letter to drop to the bottom of the folder speedily and accurately. No pinching of contents; no wrestling with guides or mutilation of folders to gain working space. Errors of misfiling are minimized.

The Drawer is Closed

Step No. 4 A light push closes the swing front, releases the Throw-Back Compressor, applies compression to the records and the drawer glides back into the case, safely latched. The filing has been done. These four simple and effortless operations not only reduce the time factor in filing and finding, but also reduce the mental and physical fatigue of the operator — minimize errors — speed record handling — preserve records and indexing in the least space — reduce costs.

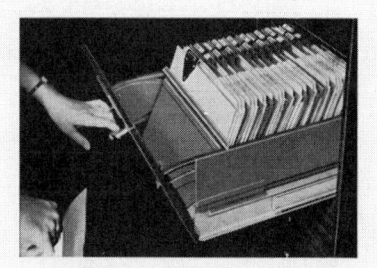

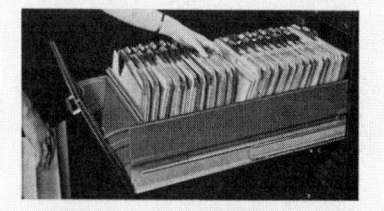

[5]

Figure 5.8. This advertisement explaining how to use General Fireproofing's Super-Filer shows only the arms of the female file clerk in illustrating how the cabinet keeps paper vertical to allow easier access. By removing women's bodies, such advertisements supported the concept of filing as machine work—work that did not require thought. General Fireproofing, *Metal Business Furniture* (1943), Trade Catalog Collection, Hagley Museum and Library, Wilmington, Delaware.

EFFICIENT

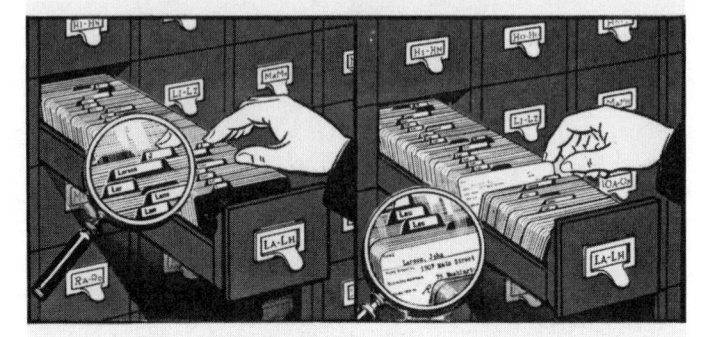

2. "John Larson" is wanted. The eye sees the guide before the hand reaches it. A *touch* of the finger—the index guide and all cards in front of it tip forward.

3. Now look! A perfect "V" opening. *TOP FINGERING* of a few cards (absolutely uniform in size) and the "John Larson" record is instantly visible—100%.

IN-EFFICIENT

2. The group of cards is spread out with two hands and the *search* is on.

3. Handling the cards one by one and *finally* "John Larson" is located.

Figure 5.9. The attempt to promote the workings of a file drawer focused on the interior, which often meant hands were separated from bodies in illustrations. This page from a Shaw-Walker office equipment catalog demonstrates how cabinet logic worked in the storage and retrieval of index cards. The appearance of male hands here shows that in practice the feminization of filing, caught up in struggles over the reorganization of office work, did not happen instantaneously. Courtesy of Grand Rapids History and Special Collection, Archives, Grand Rapids Public Library, Grand Rapids, Michigan.

disembodied hands retrieving index cards, a Shaw-Walker catalog explained how standardization and indexing ensured that its cabinet worked "as swiftly and smoothly as a fine machine." The Shaw-Walker cabinet was efficient because its consistent indexing and standardized cards determined the act of grasping paper and information; the file drawer coordinated eyes, hands, and fingers: "The eye sees the guide before the hand reaches it."[58] The filing cabinet, as a machine, presented information at a glance. The text acknowledged the hands were connected to a body, or at least an eye, but not a mind. However, when the mind of the file clerk was invoked, it was understood to perform manual work akin to the work performed by the hand. In the words of a how-to-file author: "Mind, eye, and hand can soon be trained so that they automatically act together and do team work that is invaluable."[59] As the Shaw-Walker catalog suggested, filing equipment, conceived as an automatic machine, provided the training to coordinate the senses as a team suited to repetitious and routine work. The mind was a member of this team because, as Harry Braverman argues, to the extent clerical "work is still performed in the brain . . . the brain is used as the equivalent of the hand of the detail worker in production, grasping and releasing a single piece of 'data' over and over again."[60] Tabbed guide cards and manila folders broke thinking down into a set of predetermined steps, much as the assembly line did with labor, so a clerk could handle information (as a standardized unit) without needing to comprehend the specific content. Disembodied hands illustrated that the work of the brain had become manual work; remembering had become an act of physical dexterity.

The emergence of clerical work as machine work, of the primacy of the hand over the brain, was part of a larger change in which "hands and machines began to work together in new ways that challenged previous definitions of 'manual' labor, 'craft,' and touch."[61] In many instances machines replaced physical strength, but not mental activity. However, as Rachel Plotnick shows in her innovative history of push buttons, a "popular rhetoric of simplicity, effortlessness, and no requisite skills on the part of users" structured the representations at the turn of the twentieth century regardless of what activities were delegated to machines.[62] A commonsense association between a woman's hands and dexterity meant machines that could be operated by a mere touch were machines that should be operated by women; similar to the Shaw-Walker "Built Like a Skyscraper" campaign, advertisements for push-button technology often showed young girls operating machines.[63]

In constructing many of the new specialized machine-based tasks of twentieth-century clerical work as feminine occupations, advocates sought to naturalize the changes taking place in the office by invoking as common sense the association between a woman's hands and dexterity.[64] By 1943, Evelyn Steele, editorial director of Vocational Guidance Research, could confidently note, "It is generally agreed that women do well at painstaking, tedious work requiring patience and dexterity of the hands. The actual fact that women's fingers are more slender than men's makes a difference."[65] This idea was usually linked to leisure activities and work. For the former, it was often noted that socially acceptable leisure practices, reframed as work, provided ways for women to enhance their natural dexterity and thus gain informal training in filing. As one manager commented, "I often ask a girl if she plays the piano, or if she knits, crochets, sews, or does another type of work that would enable her to acquire speed with her fingers."[66]

When filing experts invoked work outside the home to justify women's suitability for filing, they turned to the recent history of women's repetitive labor in light manufacturing. The rationalization of work that brought women into factories began in the mid-nineteenth-century textile and papermaking industries. However, as Judith McGaw argues, this work also exploited women's domestic practices. At the same time women were being assigned to repetitive labor (and to the smallest and least efficient machines) outside the home, cheap commercial goods were modifying domestic work, turning it into a series of repetitive tasks that demanded precise movements and visual discrimination. Similar to the relationship between knitting and filing, women's domestic work was understood to provide training in useful manufacturing skills. Because this informal training took place in the private space of the home, it contributed to the belief that women's factory work did not require any skills. In comparison, on-the-job training and formal education made the skills associated with men's work visible in the public space of the factory. In this way the doctrine of separate spheres, of a masculine public world of work and a feminine private world of the home, served to supply the industrializing economy with cheap and essential female labor.[67]

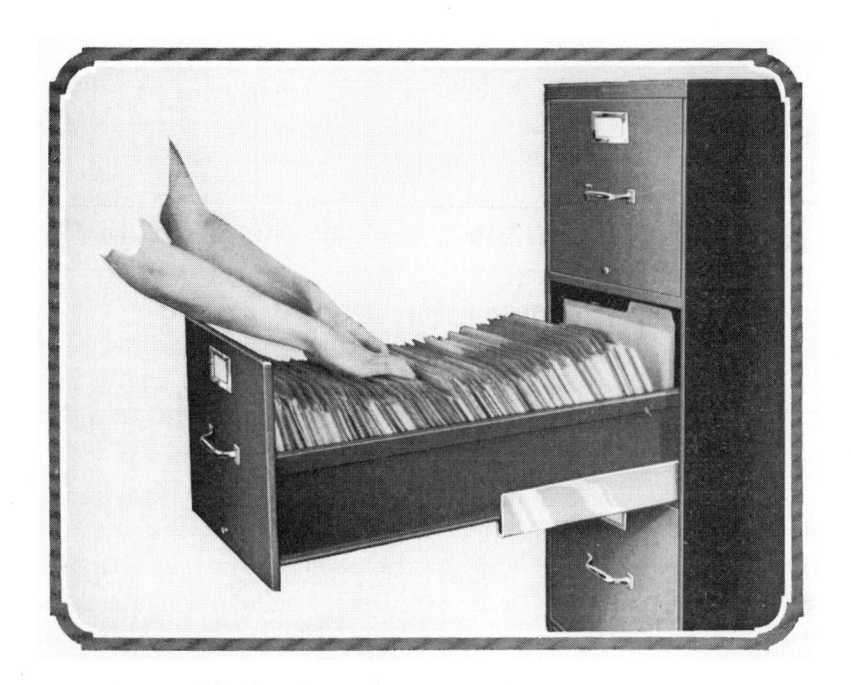

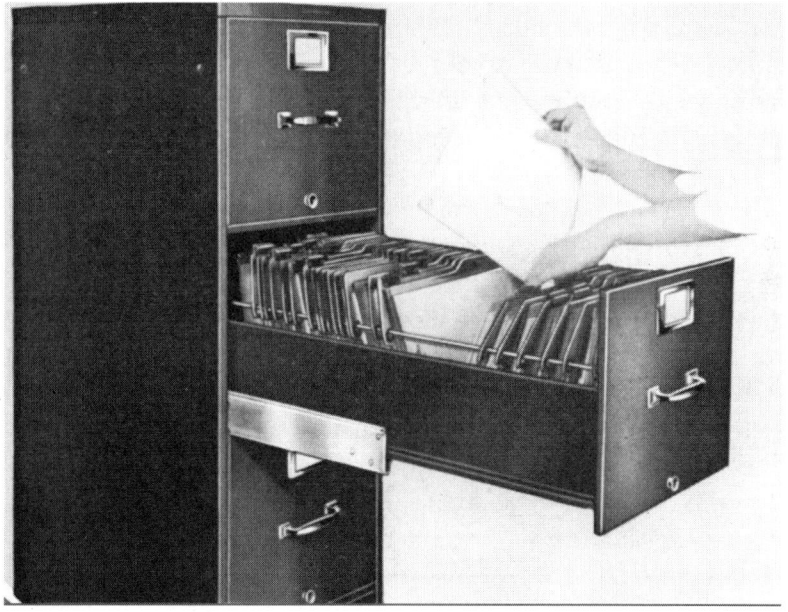

Figure 5.10. In promotional illustrations, the disembodied hands and arms of file clerks were usually identifiably female. In these images from a 1931 Globe-Wernicke catalog, the crude erasure of a woman's body emphasizes the importance of a woman's hands to the act of filing. Globe-Wernicke, *Steel Filing Equipment* (1931), Trade Catalog Collection, Hagley Museum and Library, Wilmington, Delaware.

Therefore, in paper mills, before and after mechanization, women were assigned jobs that were "monotonous and interruptible, requiring neither long training nor initiative," a common trend throughout the initial industrialization of the United States.[68] The rag room became a central site for female workers, who utilized "manual manipulation and visual discrimination" to sort rags and cut them to a standard size.[69] This anticipated the information labor of filing at the turn of the twentieth century, when standard-sized paper circulated in offices as correspondence and records, and women "operating" filing cabinets sorted paper at a glance to manipulate information.

When women arrived in the office, the historical association between femininity and dexterity, and the manual work that accompanied it, met a decades-long unease about clerical work, masculinity, and manual work. Despite a lingering belief that clerical work offered men an informal business apprenticeship, the reality was that many nineteenth-century clerks did a form of manual work. As new markets developed to meet the demands of industrial capitalism, large numbers of merchant clerks were hired to "serve as the fingers of the invisible hand."[70] Much of the work was routinized and boring for the minds to which those fingers were attached. In the words of one frustrated clerk: "Ho! The torn coat sleeve to the table. The steel pen to the ink. Write! Write! Be it truth or fable. Words! Words! Clerks never think."[71] Although their brains did not register this time in the office as work, their bodies definitely did. Michael Zakim's analysis of "desk diseases" shows this was manual work. Clerks' bodies reacted with stomach ailments, headaches, back pain, and aching hands. To ensure that clerks would be productive workers, they were encouraged to focus "on matters of diet, evacuations, exercise, air, sleep, and passions." This solution drew from the increasingly popular idea of the "self," which some writers and physicians used to argue that a man should become "a conscious object of his own observation and intervention."[72]

The "self-consciousness" of clerks fostered an anxiety about the status of their work—it was not "brain work" that would lead to the accumulation of wealth, and although it degraded their bodies it did not enhance their masculinity, unlike the strength-based manual work of an agrarian economy. The arrival of "machines" heightened this anxiety. Machines threatened to increase the visibility of clerical work as routine work and further lower its status. However, while the introduction of machines (and efficiency) did make clerical work more routine, dominant understandings of gender made it routine work that men should not do.[73] The operation of typewriters, telephones, automatic punch tabulating machines, bookkeeping machines, addressing machines, and filing cabinets became women's work. Therefore, the arrival of machines enhanced the status of male clerical work, which usually included quasi-managerial functions that tied it to the main purpose of the office, the work an executive would do at his "efficiency desk" while reaching for a file. But further down the office hierarchy, men still did the clerical work that most closely approximated a more traditional definition of manual labor, usually as shipping or mail clerks.

Although the articulation of women and machines enhanced the status of most male office workers, that did not prevent the change from fueling concerns that office work was not a truly masculine occupation—the arrival of women threatened to feminize the office and all the work done in it. The lingering anxiety that male office workers felt is evident in the depictions of men using filing cabinets as exercise equipment in Shaw-Walker's "Built Like a Skyscraper" campaign, which began in 1913 and ran for more than decade (Figure I.5). Although dressed in suits, these athletic male bodies did not belong to men who filed; the campaign's image of a young girl opening a file drawer by pulling on a thread made it clear who filed.

It is important to note, however, that men still occasionally appeared as file clerks in advertisements and textbook illustrations into the mid-1920s; sometimes disembodied male hands illustrated the efficiency of filing cabinets. That is, the changes that led to the coding of low-level office work as feminine (as information labor) were neither instantaneous nor comprehensive. Rather, while these ideas shaped the office and determined the value and status of different office tasks, for a number of reasons they were never universally followed in practice. The lingering presence of nineteenth-century office organization and attitudes meant that men filed papers into the early twentieth century. Their representation as file clerks in advertising and textbooks seems to speak to the memories (or nostalgia) of those who created the advertisements or wrote about filing. However, because the advertisements were still primarily attempts to sell the filing cabinet as a machine that required no expertise and little thought, they often contained contradictory messages.

A 1916 advertisement from Globe-Wernicke showed a man filing, with accompanying text that promoted filing cabinets through generic assertions about swiftness, noiselessness, and ease of operation (with a finger), along with claims about saving time, thought, and energy. However, these attributes did not make the filing cabinet "automatic" or the clerk efficient; the gender of the clerk centered him and made him a "brain-worker." A four-panel comic strip from 1921 offers another take on the status of filing when a man did it. After being told that he will not be getting a raise, a male file clerk quits and scatters files around the office floor. In the final panel of the cartoon, the clerk discovers his boss has given him a raise after all, because the boss has realized that he is the best file clerk the office has ever had.[74] From the mid-1920s, however, advertisements rarely represented men as file clerks. If a man appeared near a file drawer, it tended to be a card file, as the Shaw-Walker catalog page described above illustrates. The index card suggests a kind of information that is different from correspondence, numbers, and figures. Perhaps the hands belonged to a salesman retrieving information on *his* clients? This male presence indicates that although filing, as information labor, was coded as feminine, a continuing and unresolved argument remained about authority, the meaning of filing, and what it meant to be in charge of files.

In an attempt to sell the "ideal system of filing office letters," a Shaw-Walker brochure stated: "You, a stranger to the file, can't look long for a letter because everything is in full view. . . . The names of all the regular correspondents can be read the instant the drawer is pulled out. You don't need to be an expert to find a letter. Simply open the drawer."[75] Presented as a machine, a filing cabinet was easy to operate, with visibility equated with accessibility. However, while you did not need to be an expert to operate a filing cabinet, you did need to be a woman. If a stranger could operate a filing cabinet, then anybody could. "Anybody" became a specific body, a woman, via the logic that if anybody could file then only a woman should file, because a man could do other work that a woman could not do. Therefore, the "stranger" Shaw-Walker's advertisement copy

Figure 5.11. In the early twentieth century the status of filing was not fully resolved, and men occasionally appeared as file clerks in advertising and textbooks. In this 1916 Globe-Wernicke advertisement a male file clerk is identified as a "brain-worker," in contrast to how women were represented as file clerks who did not need to think. Author collection.

celebrated as being able to use a filing cabinet labeled the gendered subjectivity of a deskilled worker, a worker alien to the masculine skills used to define the office as a modern work space in the early twentieth century.

As the handling of paper became synonymous with the handling of information, it accentuated the instrumental encounter with in-

Compartment Filing
SOLVES MANY FILING PROBLEMS

Art Metal Adjustable File Supports can be inserted or removed without disturbing file contents or guide rods. No springs or gadgets are involved.

The New Adjustable File Supports

THE NEW ADJUSTABLE FILE SUPPORTS, AS DEVELOPED BY ART METAL, ARE FAST PROVING THE SIMPLEST YET MOST FAR-REACHING ADVANCE IN FILING TECHNIQUE IN RECENT YEARS . . . BY ELIMINATING TWO HAND MOVEMENTS THEY SPEED FILING OPERATIONS . . . BY DIVIDING THE FILE INTO COMPARTMENTS THEY ELIMINATE MANY FILING PROBLEMS.

Figure 5.12. The lowly status given to filing and information labor is captured in promotional materials that used disembodied arms and hands to represent the act of filing. This image from a 1940 Art Metal catalog reduces the hands to sketched outlines, further abstracting the work of filing. Courtesy of the Fenton Historical Society, Jamestown, New York.

formation that I have called information labor. An employer did not expect a low-level clerical worker to be informed as information passed through her hands. Having information at one's fingertips meant grasping paper with hands disciplined by the predetermined pathways created by a file drawer. The priority given to this more feminine form of manual labor allowed the further routinization of the office in the early twentieth century and the representation of filing as more rote and less intellectual than preceding clerical work. Information labor, the articulation of dominant ideas of efficiency, system, and gender, made women perfect file clerks who performed scientific management's ideal of "pure labor . . . uninterrupted by the coordinative and communicative demands of administration."[76]

Chapter 6

THE IDEAL FILE CLERK
Controlling Gender in the Office

In one of the first books to be published on scientific office management, New York University professor of economics Lee Galloway offered the following description of a file clerk:

> File clerks must be rapid and accurate at detail work, attentive to operations which are continually repeated but slightly varied, and must have automatic carefulness in replacing papers. The nature of the activity calls for intelligent and trained women rather than men or boys. Women can handle detail work with more persistent accuracy and patience and with swift, automatic expertness. The head of a centralized filing department, however, occupying a position coordinate with that of the other department heads, should preferably be a man.[1]

Galloway summarized the gendered division of office work and the conflation of women's work with the machinelike capacities granted to a filing cabinet, what I have argued constitutes information labor or the ideal mode of clerical work in the early twentieth century. Galloway was explicit: a female file clerk and a filing cabinet shared the same attributes. The "automatic carefulness" and "persistent accuracy" commonly attributed to the filing cabinet blurred with women's natural dexterity and their purported affection for order.[2]

While gender was a formative category in the organization of the office, it intersected with class, race, age, and sexuality to create

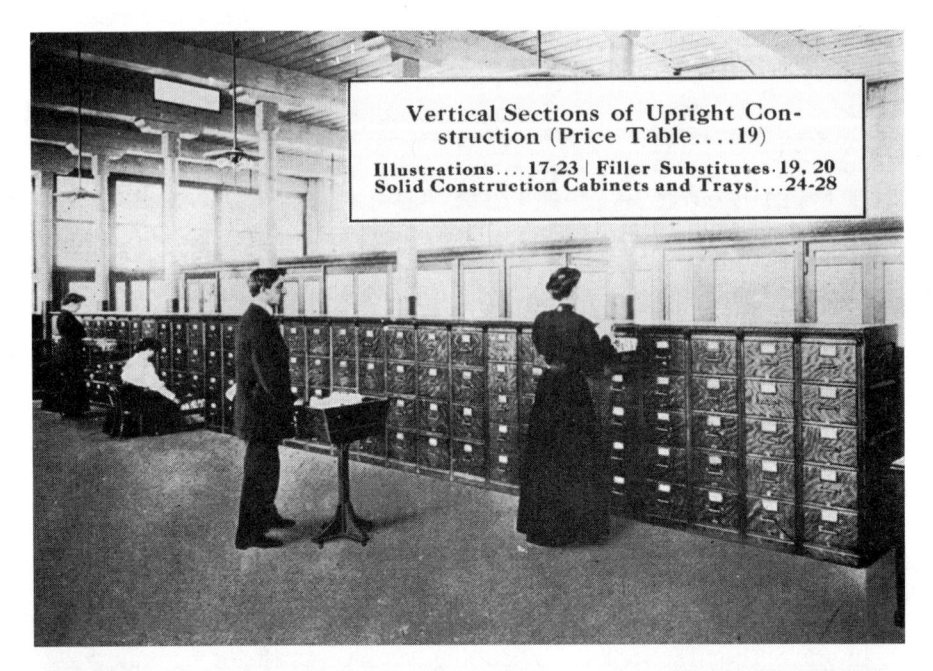

Figure 6.1. The ideal file clerk was a woman who worked to serve men and to provide men with information, as these images from early twentieth-century filing equipment catalogs illustrate. Author collection; Yawman and Erbe, *Record Filing Cabinets* (1910), Trade Catalog Collection, Hagley Museum and Library, Wilmington, Delaware.

a more complicated picture of the women thought to be ideal file clerks. Galloway's comments about "intelligent and trained women" suggest as much about his beliefs regarding class and race as they do about his attitudes toward gender. His emphasis on training and intelligence reflected not only the ideal that women clerical workers should fit into the emerging white-collar middle class but also the belief that office work would impart middle-class values (moral and social respectability, education, and upward mobility, all anchored in a heteronormative family centered on female domesticity and a male breadwinner) to working-class and immigrant workers who filed, whether they aspired to those values or not. The presence of immigrant workers and their American-born children acknowledged the racial organization of the office and the whiteness that supported it. Clerical work gave white women, broadly defined, employment opportunities that were unavailable to Latina, Black, and Asian women except on a very limited basis, usually within their own communities.[3]

Therefore, ideas associated with middle-class values emerged in office literature and career advice literature to police whiteness in the white-collar world of the office and to universalize the dominant racially organized office of corporate capitalism as *the* office. *Middle class* as a synonym for *white* also provided the scaffolding for the hierarchy in women's clerical work. White, Anglo-Saxon, middle-class women who worked in offices sought positions with responsibility, such as that of personal secretary or even head of a file department (a position with enough responsibility that Galloway believed a man should hold it). The women whose primary work involved actual filing tended to be immigrants or the American-born children of immigrants whose whiteness and middle-class identity, while more dubious, nonetheless gave them access to the office; these were the women who needed training.

In this chapter I examine how class, education, and age appeared in publications addressing the novelty of women office workers to complicate a gendered understanding of clerical work; this is made explicit by accounts of attempts to train file clerks and to get filing recognized as a profession. The chapter also offers a particular example of the anxiety and contradictions that Ben Kafka argues constitute paperwork, the anxiety generated by the actions of people tasked with doing the work that was meant to make paperwork work.[4]

File Clerks by the Numbers

Moving away from the prescriptive writing of Galloway, who were the women performing clerical work? Census data and surveys performed by the Department of Labor's Women's Bureau and religious groups in the 1920s and 1930s provide a broad answer to that question. The 1930 U.S. census recorded almost two million women as clerical workers. These women had not replaced men; rather, they represented the increased numbers of clerks and, importantly, the increased percentage of clerks who did only routine work. The blurring of clerical and white-collar work in the 1930 census meant that 51 percent of those recorded as clerical workers were men.[5]

Surveys found that the typical female office worker was a woman in her mid-twenties, likely on her third or fourth office job. On average, she earned $25 per week. In larger cities, half the women working in offices had begun office work before they turned seventeen years of age; this number was halved in smaller cities.[6] Most of these women were born in the United States. However, some surveys showed that up to two-thirds of workers were the daughters of immigrants (the proportion was as low as one-fifth in smaller cities).[7] Regardless of job title, clerks were expected to have some high school education, but there was no expectation they would have graduated.[8] Any education beyond some high school was considered a hindrance rather than an asset. As one employer said of women clerks, "Education makes them dissatisfied to do routine work."[9]

Within the limited employment opportunities available to women, clerical work provided reasonable pay and decent working conditions, if not job satisfaction.[10] Working close to forty hours a week, the average office worker encountered shorter hours and better working conditions than did her counterparts in trade and industry; the two-week paid vacation for some clerks in city offices signaled this difference. Clerical pay varied depending on the city and the type of office but consistently compared favorably to the pay for other kinds of employment open to women. As suggested by the average pay of $25 per week, the national monthly median pay for a female office worker was $99, reaching $109 in New York City.[11] In 1926 this $1,200 annual salary, while considerably less than the average worker's income of $2,010, was not significantly lower than the $1,309 average income of nonunionized manufacturing workers and the $1,275 average salary of teachers, principals, and superintendents in public schools.[12] File clerks were paid below the median, with a third earning less than $75 a month or $900 a year; one survey called the position a "beginning job," although only around 5 percent of clerical workers were designated as specialized file clerks.[13] The two largest categories of female office workers, each comprising one-third of workers, were "general clerk" and "stenographer."[14] However, general clerks almost certainly filed. Despite the focus in office management literature on large offices with specialized clerical positions, most women worked in small offices with fewer than twenty-five employees, where employers expected all female clerical workers to file, regardless of their job titles.[15]

Age

One in four women clerical workers was older than thirty years old. The archetypal "file girl" or "white-collar girl," the female clerical worker who appeared in articles, books, and popular culture, represented the 50 percent of women office workers who were younger than twenty-five years of age.[16] These younger clerks were at the center of the debates that occurred when women began to work in offices alongside men (young, old, married, unmarried, but all assumed to be heterosexual). While concerns about the "mixing of the sexes" had largely abated by 1920, the office remained a space outside the home that gave rise to debates around sexual norms, ideals, and meanings, issues that were often addressed via representations in films and short stories.[17]

The intended humor in many of these popular portrayals focused on women as sexual objects. These representations functioned to negate concerns that the masculine space of the office would threaten gender norms and make women less feminine. The stylish, modern young woman office worker in popular culture could also offer alternative conceptions of femininity and work. However, films presented such a woman as someone who wore her fashionable clothes more comfortably than her independence. Her search for a husband calmed male anxiety by showing that even a modern woman still needed male attention. She remained dependent on such attention, and any changes in her behavior were ultimately not significant enough to warrant concern.[18] Critically, Lisa Fine argues,

IN THE FINEST PRIVATE OFFICES

Figure 6.2. Women were frequently featured in promotional materials modeling how to operate filing cabinets, but as they did so, they were also being presented as models for the assumed male gaze of the reader. In this page from a 1917 Wabash Cabinet catalog, the male worker models the sexualized gaze of the male reader while the file clerk avoids eye contact with him. Wabash Cabinet Co., *Wabash Filing Cabinets* (1917), Trade Catalog Collection, Hagley Museum and Library, Wilmington, Delaware.

films represented the young office girl as, at her core, a "good girl." Yes, she was not a completely passive, pure, and chaste nineteenth-century woman, but she was definitely not a promiscuous flapper. In a world of change, representations of office girls frequently provided reassurance that goodness still existed and the "natural" gender order remained intact despite the fact that the modern world now included young women who handed files to older men.[19]

As a poet writing under the pseudonym Better Service put it in a 1918 issue of a magazine devoted to filing:

> Be-hold the little File Girl, so sweet and de-bon-air,
> With noth-ing on her lit-tle mind except her fluff-y hair.
> Her dis-pos-it-ion's splen-did—see how she smiles at me!
> The while she puts File ten-o-four in Folder ten-o-three.[20]

Better Service was correct to suggest that the youth of the file girl became an issue when the promise of automatic filing failed to deliver. As noted in chapter 5, the fault for misfiling was always laid at the hands of the file clerk, not the system. As a form of misuse and error, misfiling is an example of what Victoria Olwell calls "bodily malfunction," in the sense that it punctures the fiction that a filing cabinet is automatic. It brings attention to the fact that filing is a product of someone's labor, not an abstract system. As Olwell argues, "The cloak of invisibility covering the body drops away . . . the moment that body makes a mistake."[21] In this case, the mistake makes visible a very particular body (and mind, albeit a distracted one), that of a young woman. Age and sexuality trumped gender to provide an explanation for the failure of the innate dexterity and sense of order usually invoked to present women as ideal file clerks. Therefore, the occasional discussion of misfiling highlighted a lingering anxiety surrounding the presence of young unmarried women in offices but articulated it to, and through, the limitations of the filing cabinet as a machine. These ideas became difficult to disentangle because they both expressed concerns about productivity grounded in the limitation and contradictions inherent in dominant ideals about machines, labor, and gender; this is also an aspect of the contradictions of paperwork, often experienced as carelessness, which Kafka isolates in his analysis of bureaucracy.[22]

The idea of the "file girl" as a distracted young woman working in the city, who actually did have things "on her lit-tle mind" other than her "fluff-y hair," just not work, fits into the office as a space where debates arose concerning sexual norms, ideals, and meanings. In this scenario, misfiling provided an opportunity to underscore the belief that young women were obsessed with finding husbands ("see how she smiles at me!"); the anxiety of single women (and men) was conflated with anxiety over paperwork. An experienced female office manager captured the concern about the priority young women gave to socializing in the city when she instructed file clerks to "leave fine clothes, the theatre, pleasant parties, Tom, Dick and Harry, at home." This was necessary because "important tasks cannot be accomplished with your hands while unimportant details fill your head. You cannot file 'Amusement' under 'Work.' They are at the extremes of the alphabet."[23]

The author of one of the most detailed 1920s surveys of women office workers concluded that young female workers did possess a "sublime faith in early marriage," and she recommended that they be given "vocational training for marriage."[24] While the survey showed that a young woman office worker was likely also concerned with her salary (and her "own personality"), marriage was the only union in which she was interested.[25] Faith in marriage encouraged young women's acceptance of the short-term nature of work, which the survey author suggested was one reason for the lack of unionization in clerical work.

Despite worries about young unmarried women, "Mrs. File" was not offered as a solution to the concern that "Miss File" might misfile. Surveys showed that more than 80 percent of women clerks were unmarried. The proportion was assumed to be higher in what a federal government survey referred to as the "blind-alley class" of routine work, which included filing.[26] Although women often left office work when they got married, it is unclear whether a formal "marriage bar" existed. In larger firms, especially in banking and insurance, where it was more likely that rules required a woman to resign when she got married, the resignation was nevertheless commonly considered to be her choice.[27]

Even in the absence of a formal marriage bar, an "assumed heteronormativity" played a critical role in determining the kinds of jobs available to young women and in justifying the lower wages paid to women.[28] Significantly, wages were tied to the idea of a family home with a man as the main breadwinner; any income a woman brought in was considered a supplement. This notion, along with the informal marriage bar, contributed to assumptions that women were short-term workers and associated them with jobs that required limited training and skill and offered little opportunity for advancement—efficiency depended on gender and sexuality to lower labor costs.

Class and "Personality"

Class and ethnicity offered solutions to anxiety about misfiling and misbehavior and therefore clarified which young women were best suited to filing. Career advice writers deployed the word *personality* as a synonym for *control,* and their calls for control positioned the routines of low-level clerical work as useful for teaching young women discipline in the form of the manners and norms the writers associated with middle-class identity. The focus on class connected debates about women clerical workers to an increasing awareness that a middle-class identity existed, although it was an identity that still needed to be staked out.

In the nineteenth century, claims for middle-class identity came out of changes in the character of work and work environments, especially in the office.[29] These included the separation of factory and office and increased attention to cleanliness and refinement. Although what became known as white-collar work was critical to assertions about middle-class identity, the expansion and specialization of office work did not always fit comfortably within the separation of nonmanual and manual work, of "headworkers" and "handworkers," that laid the groundwork for this new class identity. The more manual forms of clerical work brought into the office a number of workers who did not fit into the developing definition of middle-class identity.

Conflating white-collar work and middle-class values, advice literature presented "personality" as something low-level clerical workers should aspire to if they were working in middle-class spaces. Personality could be gained through exposure to the technologies used in office work; the training of low-level clerical workers was directed toward "how to operate the self" as much as it was concerned with the operation of the particular machines they would use.[30] Filing required, and through repetition could teach, what a senior female office manager identified as the "three necessities" of efficiency: concentration, accuracy, and "good nature." Introducing another office technology, she explained concentration as the need to "control your thoughts as you must your pencil, with a firm grip despite outside disturbance and inward annoyance."[31] The assumption was that a clerk would misfile if she failed to compartmentalize, if she failed to keep her work duties and her personal concerns ("fine clothes, the theatre, pleasant parties, Tom, Dick and Harry") in their proper place and order—an explanatory structure that evoked the act of misfiling.[32]

At the lower end of the clerical hierarchy the conflation of middle-class values and efficiency was not equivalent to the collapsing of identity into work that made professions the basis of middle-class identity. Rather, middle-class was an aspirational identity for low-level clerical workers, making the ideal worker a woman whose job was defined by the rationality of the machine she operated, a worker defined by and subsumed within routine to ensure the smooth functioning of the system.[33] Therefore, in the conflation of middle-class values and efficiency, personality became a way to be neutral, to be not noticed. The aim, according to a former secretary, was to prevent "unbridled individuality running riot all over the place."[34] In the office the "individuality" of women of dubious class and ethnic background marked deviance from acceptable middle-class behavior; in other contexts for different people, individuality was an acceptable part of middle-class identity.

Clothes became a site of struggle over control and individuality in the office; as noted above, women workers might be told to "leave fine clothes . . . at home."[35] But the importance that female clerical workers assigned to clothes is evident in records of their expenses, which show that for most, the costs related to their appearance were second only to costs for room and board.[36] This response to the expectations placed on women by the middle-class space of the office highlights the unpaid work that women had to do to maintain the appearance demanded by a middle-class work environment. Judith McCaw shows the importance of this unpaid work, which included maintaining the appearance of the male clothing that signified the middle class. A white collar had to be kept white,

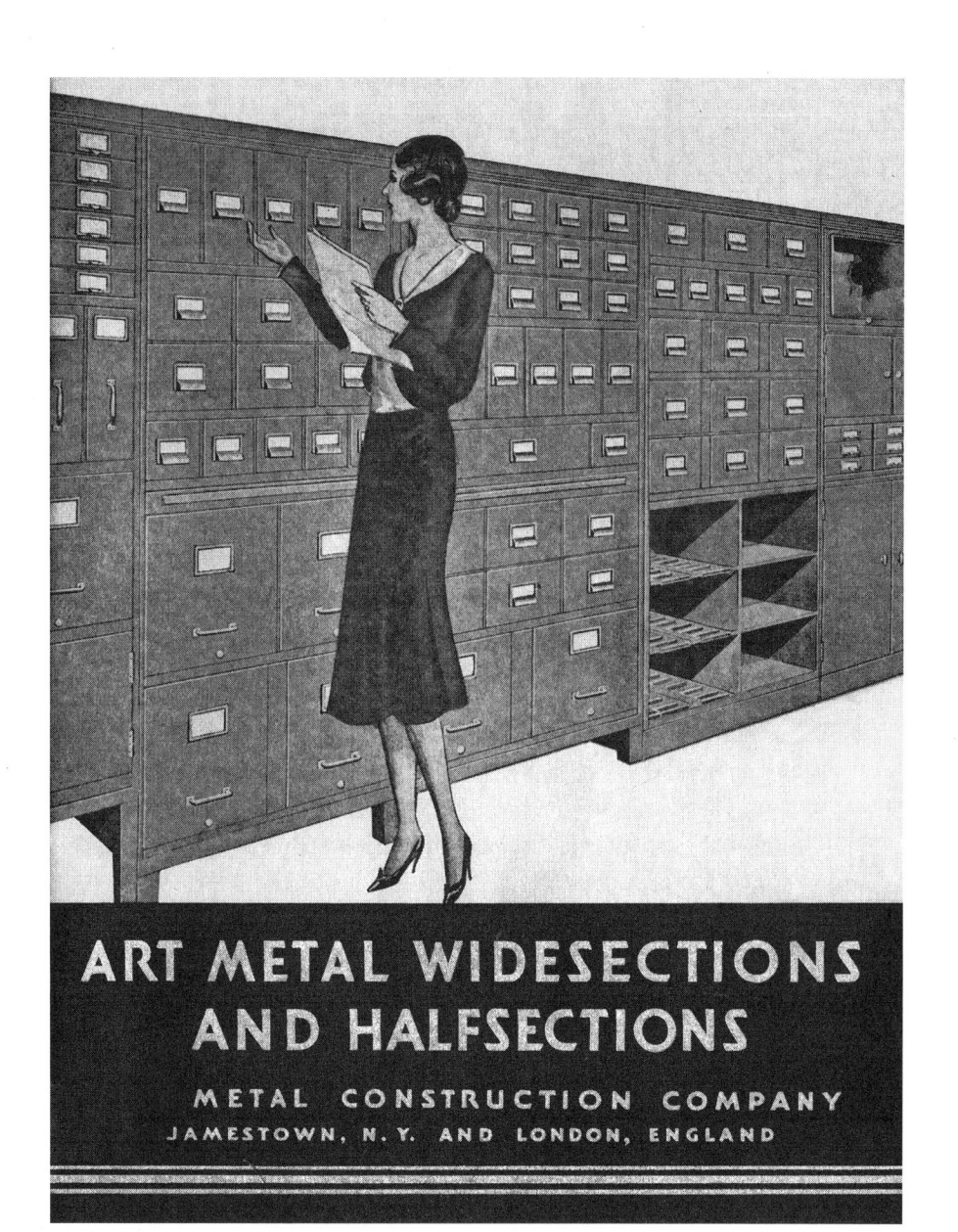

Figure 6.3. Although office guides argued that women workers should ignore fashion and dress practically for the office, the idealized representations of office work in promotional literature tended to present file clerks as women attuned to fashion trends. This image from a catalog cover also illustrates the importance of representing women as able to access tall cabinets, usually assisted by fashionable high heels as well as the artist's sense of proportion and scale. Art Metal, *Widesections and Halfsections* (1931), Trade Catalog Collection, Hagley Museum and Library, Wilmington, Delaware.

because that whiteness assured coworkers and customers that the person they were dealing with was respectable. By the 1920s, with the disappearance of detachable collars, this requirement meant that "someone had to keep the whole garment white."[37] However, by that time that someone was likely to be a woman employed outside the home in a commercial laundry.

For women who worked in offices, respectability was measured by how they used clothes to deal with the conflicting ideas of fashion, sexuality, and morality. Women's struggles over what to wear figured into debates about the proximity and behavior of men and women in offices. As fashions changed and young clerks navigated the contradictory signals they received about sexuality in the office, debates over "transparent flimsies in blouse and hose, and the high-healed [sic] unhygienic pumps" positioned younger women against older women.[38] Advice literature, usually written by older women, focused on curbing female desire in the name of middle-class behavior, while accepting that marriage dominated the thoughts of young female clerks. Unwanted sexual advances and harassment from men, however—although common and a concern for some women—were not behaviors to be managed or discussed publicly.[39] Women had to curb their emotions and the "feminine instinct to attract, to awaken a response" in a space where they were hired to "add to the general attractiveness of the office," as the male manager of an employment service put it.[40]

In response to such issues, a call for an unofficial uniform had accompanied the hiring of women as clerical workers. Advocates also believed that a uniform would signal to women that their social status did not have a place in the office; women could also use a uniform to hide their social status from other workers.[41] In the early 1910s, the Metropolitan Life Insurance Company strongly encouraged its women clerks to wear an unofficial uniform consisting of a long dark skirt covered with a white apron, a white blouse, and a black bowtie.[42] But in most offices, dress codes were little more than series of suggestions regarding styles considered appropriate and inappropriate. By the late 1930s, the Transcription Supervisors Association told high school girls interested in office work that it was acceptable to dress in ways that were "modish and becoming but not rakish or bizarre." They were also informed that "unobtrusive jewelry and straight heels on polished shoes are recommended and fragmentary heels and toes are . . . akin [to] poor taste."[43]

The articulation of personality and practicality produced specific clothing suggestions for women who filed regularly. They were discouraged from wearing the "tight shirt-waist and skirt" then popular among office workers. Responding to the need for clothing that did not restrict arm movement, a booklet targeting potential file clerks suggested that "the loose Russian blouse or Norfolk jacket worn outside the skirt is neat and comfortable, and does not expose you to the possibility of its working loose untidily, as sometimes happens with the best-behaved of shirt-waists."[44] One newspaper columnist went so far as to attribute the demise of the corset to the "loose-fitting sensible style" of "business women whose office day usually consists in bending over a desk, playing the typewriter keys or stooping over filing cabinets."[45]

Teaching Filing

Education offered a more explicit attempt to prevent young women from misfiling. By the 1930s, most office managers expected lower-level clerical workers who filed or did billing and payroll to have a junior high school education, preferably with some specialist training in typing, stenography, or filing.[46] This expectation came out of significant changes in the high school curriculum in the first decades of the century. Discussions about the social role of education began at the end of the nineteenth century, when less than 5 percent of American children ages fourteen to seventeen attended high school. In an attempt to encourage students to stay in school past the eighth grade, public high schools began offering commercial and vocational courses, previously the exclusive purview of small private schools; by 1940, almost 75 percent of fourteen-to-seventeen-year-olds attended high school.[47]

While more children attended high school, many public school educators were concerned about the changed expectations that commercial education brought with it. Educators generally had expected that after students were lured into high school by promises of commercial education, most of their classes would be in more traditional subjects and directed toward the goals of character building and citizenship training.[48] Business leaders, however, promoted commercial and vocational education in the public schools because they wanted government to take responsibility for better preparing youth for twentieth-century jobs. Although traditional subjects remained the cornerstone of high school education, public education increasingly came to be understood in terms of the needs of the economy, not those of society at large. Courses with names such as "Business English" and "Industrial History" emerged during this era as humanities-based education was forced to adjust to changing social and economic expectations.[49]

By the 1930s, the increasing demand for office workers to possess "personality" encompassed what many employers valued in formal education as much as skill sets for particular jobs. Personality, defined as middle-class deportment and behavior traits, replaced "character" as one of the main goals of education. A move away from character to personality was consistent with the early twentieth-century ideas of self-improvement evident in advertising and "success" literature. In this wider shift, a focus on outer appearance and behavior displaced the prior focus on the moral and mental qualities central to character. However, when personality trickled down from managers to clerks in the office, it was not about winning friends and influencing people. It was about fitting into a middle-class space; it offered a limited mode of self-improvement that did not celebrate individuality or the world of entrepreneurship. A pamphlet distributed at a New York high school listed the important employment skills that students would gain if they stayed through the end of high school: "to write and speak correctly, to be wide-awake and well informed, to be able to understand and follow instructions exactly." "These are things," the text emphasized, that "a school cannot teach in one or a few years."[50] This was an era in which *good* and *efficient* increasingly came to mean the same thing, and that meaning was increasingly linked to middle-class values.[51]

The perceived necessity to foster a middle-class (white-collar) sensibility was a response to the kinds of girls who ended up in commercial education tracks and then in offices. In high schools, the students in commercial education were a subpopulation rather than a cross section of the schools' student bodies. While many high school students actively chose commercial courses, many teachers viewed these courses as "dumping grounds" for students who struggled in traditional academic courses. These students were disproportionately female, were identified as poor academic achievers, and came from a lower social class than other students.[52]

Filing emerged as a subject in this new area of commercial education taught in high schools, business colleges, and small private schools run by office equipment companies. In a teachers' guide to a filing textbook published by office equipment company Yawman and Erbe, the author argued that filing was a critical skill for managing "the great mass of detail that comes to the average office man's desk under modern conditions"; learning to file was important because it would "encourage the student in orderliness, in system, and in carefully regulated handling of routine."[53] Routine and system would discipline mind and body, helping a student to develop the "personality" to work efficiently in an office.

Filing fit within the gendered articulation of mental and manual labor that structured commercial education. Boys took courses to prepare them for work in administration and management, and girls took courses in stenography, typing, and filing. Textbooks and lesson plans explicitly and implicitly presented filing in a way that showed how thought in clerical work was defined in terms usually associated with manual work. Speed was critical to the conception of "brain work" in clerical work, which meant any thought involved needed to be simplified to replicate the actions in factory work. Within the specifics of teaching filing and the broader context of commercial education, an assembly-line thought process made filing an instance of information labor.

To meet the demand for commercial education, a handful of office equipment companies produced textbooks and equipment for teaching filing in classrooms. These classrooms were not limited to public high schools (or the evening schools most high schools ran)—they were also found in parochial and other private high schools, vocational schools, continuation schools, normal colleges, colleges, business colleges, and commercial schools. The salesmen assigned to sell teaching equipment prized private and Catholic schools in particular because they offered the possibility of direct sales without the complications associated with gaining approval from local school boards.[54]

Filing was rarely taught as a stand-alone course. The teaching kits were intended for filing units in the two-year and four-year commercial courses offered in most high schools by the 1920s. At Lowell High School in Lowell, Massachusetts, about 45 percent of students took commercial courses. The school's Office Practice Department had forty-three drop-cabinet oak desks, each with a typewriter. The classrooms also contained eleven calculating machines, one mimeograph, one multigraph, four adding machines, one bookkeeping machine, check protectors, and filing cabinets. The relative importance of filing as compared with typing is evident in the resources devoted to the school's Stenographic Department, which had more than 150 typewriters and five teachers.[55]

In 1923, Lowell High School started using Library Bureau equipment to teach filing. The company produced special "miniature filing equipment" to allow students to practice filing at their seats, instead of only reading about it or having to wait in line to practice briefly on one or two full-size filing cabinets in the classroom. In

Figure 6.4. This 1923 promotional photograph shows students at Lowell High School in Massachusetts practicing filing with equipment produced by the Library Bureau. The company introduced miniature filing equipment to allow high school students to practice filing at their own desks instead of waiting in line to use full-size cabinets. Author collection.

three years, the Library Bureau made more than $90,000 in sales of teaching equipment to more than 250 schools.[56] Sitting at tables, students used small rectangular boxes, advertised by the Library Bureau as miniature replica drawers, to file 4-by-6-inch pieces of paper that were scaled-down copies of incoming and outgoing letters and carbon copies. The miniature file box and seventy-five letters were part of a practice kit that included forty-two guide cards (the number usually housed in a standard-size file drawer), the *A* and *B* sections from a 320-division alphabetical index, and state and town guides from Illinois and New York. In addition, the supplies included two hundred plain cards, twenty-five cross-reference sheets, twenty-five numeric cards, ten card samples, and six catalog cards.[57]

To teach "carefully regulated handling of routine," the lessons in a filing course focused on what was usually called classification, but could also be called indexing, following the regular blurring of the two concepts in filing literature.[58] However, students were trained to use a classification system, not to create one. In longer courses, a student used the same set of correspondence to practice different types of alphabetical systems, a geographical system, or, very occasionally, a subject system. Library Bureau salesmen pitched the idea that "by treating this correspondence from various angles the student develops a sense of classification and is made to realize the importance of accuracy."[59] According to this argument, classification, in its prioritization of rationality and procedure, was the gateway to understanding system as central to the organization of work and production.

Teaching the classification of records and where to place them in a file drawer also introduced students to an instrumental encounter with information. One filing expert informed his readers: "The slowest part of the operation should be decision. Time enough should be taken to decide without a doubt the proper placing of the paper. The rest should be mechanical and swift."[60] While the application of the classification system should apparently be the slowest part of filing, it should not be slow. The intention was to classify rapidly by reading only particular parts of a document. Because it was not necessary to understand all aspects of a document, any thought required could be as automatic and swift as the handling of papers.

To operationalize reading in the interests of efficiency, the student was trained to look quickly and carefully at the "name on the

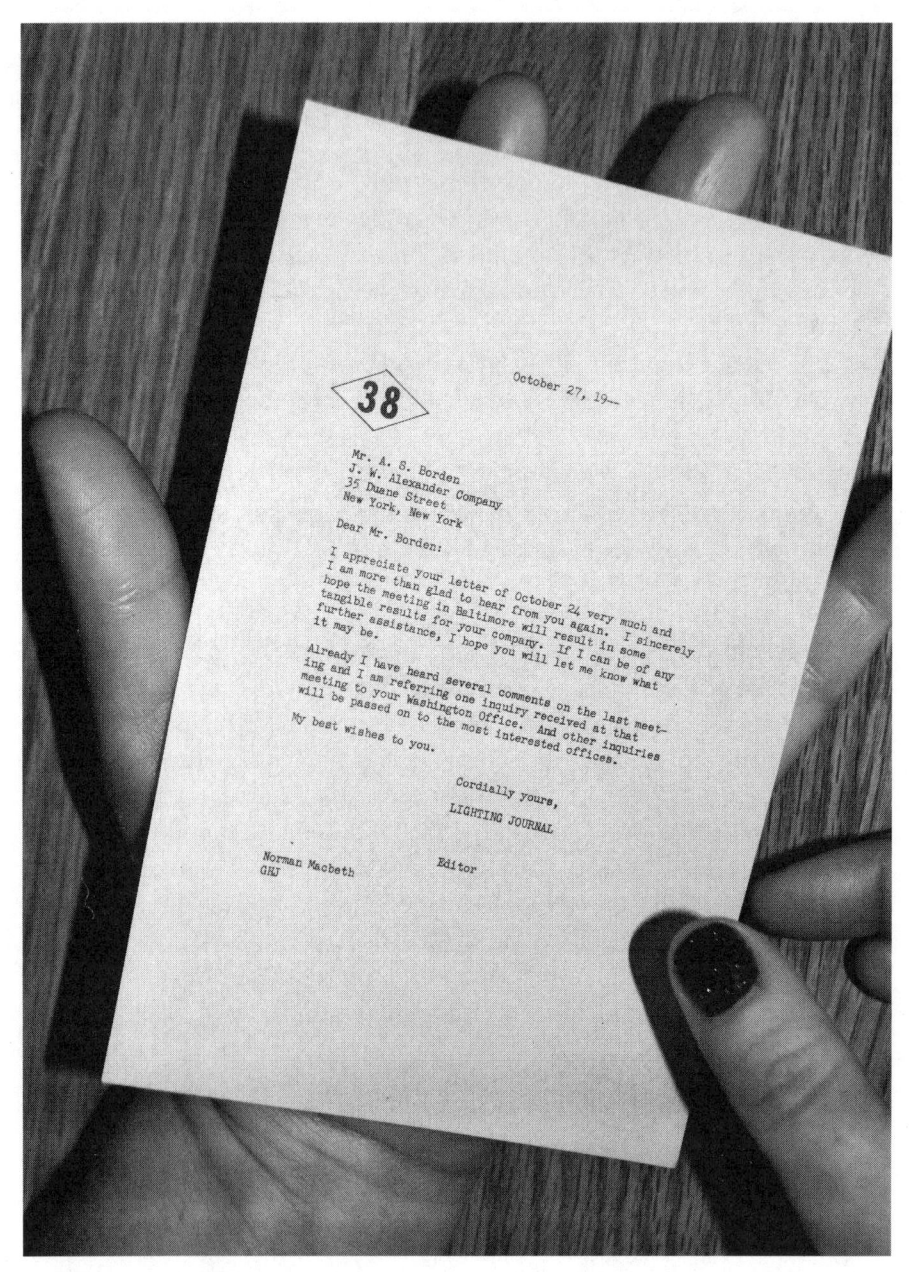

Figure 6.5. The Library Bureau classroom equipment was based on reproductions of letters purportedly taken from a real business, which may account for the spelling errors that frustrated teachers. Photograph by author.

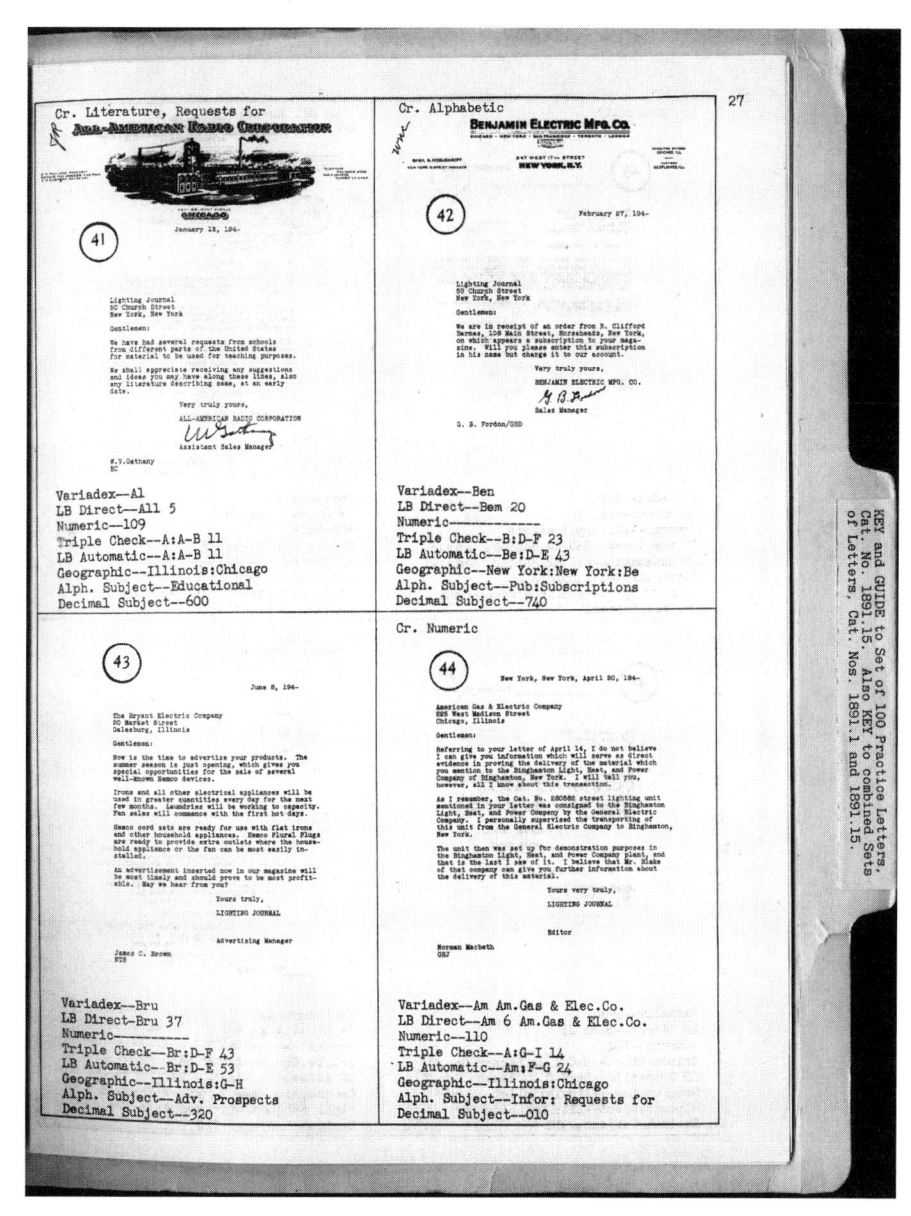

Figure 6.6. The letters in the Library Bureau's classroom set were designed so that students could use them to practice filing in multiple different systems. This page from a teachers' guide has the correct coding for different types of indexes, including alphanumeric indexing and subject indexing. Author collection.

letterhead, name of the party addressed, name of the party signing the letter, name of the subject or person mentioned in the letter."[61] Having located these names, the student was instructed to prioritize the "relation of the letters" to determine where the correspondence should be placed. This involved a focus on the precise nature of the relation of letters, as when someone might use a "dictionary, directories or something similar" rather than "read by sight."[62] In this way a clerk was encouraged to see things granularly, to reduce what needed to be comprehended. Alphabetical indexes that utilized hundreds of subdivisions consisting of the first two or three letters of a name provided another way to reduce the possibility a clerk would misfile a document by limiting the scope for interpretation.

The primary mode of encountering information that the teaching of filing sought to instill is evident in the problems associated with teaching subject filing. To identify a subject, a clerk had to do more than simply use "dictionary mode" to locate a sender's name or location within a piece of correspondence. But that "more" the subject category identified could vary depending on who read the document. The male head of a large filing department argued that "the personal equation" involved in subject filing opened the system up to inconsistencies, making filing an "expression of brain work."[63] In contrast to the ideals of automatic filing, "the subject file is not an attempt to eliminate brains, to make letters put themselves away."[64] According to a 1921 filing textbook, the problem was that the subject file introduced "the vagaries of the human mind." As a result, "the same mind frequently approaches the same subject in different ways," and the approaches of two minds differ "with still greater frequency."[65] In subject filing, classification, the order of the alphabet or numbers, could not be used to discipline workers, to remove the discretion of individual clerks. The solution to this problem, such as it was, involved training people to understand the business as a whole in the hope that this knowledge would lead to consistency in subject filing; this was the very knowledge that the filing cabinet was meant to eliminate, so a "stranger" could easily find information. The need for this additional knowledge meant that subject filing was rarely, if ever, taught in short filing courses.[66]

Filing as a Profession

The skills associated with subject filing bolstered the attempt to get filing recognized as a profession.[67] The women who wanted to be part of this profession were unquestionably middle-class; they did not need to be taught personality, they needed to learn the science of indexing. That is, a woman who succeeded in filing took pride in her work and possessed ample amounts of curiosity and ambition. This was a woman who closely identified with her work, such that self-improvement and work training became indistinguishable—that is, she claimed a middle-class identity. This produced an understanding of filing that removed it from the associations of information labor; hence, the movement to professionalize filing sheds light on the boundaries of information labor and which particular women were considered most suited to operate filing cabinets.

In 1919, the Library Bureau published a forty-nine-page booklet titled *Filing as a Profession for Women*.[68] The company distributed the booklet through its schools and salesmen and also sent it to young women who responded to Library Bureau advertisements about filing; on occasion, it was used as a textbook in filing courses, including a course at the University of Vermont.[69] In an attempt to entice women into filing, the booklet took the rhetorical stance of claiming that filing was a profession. Therefore, it described a world where most firms had "file executives" as well as file clerks. Although a file executive was at the center of the claim that filing was a profession, the position of file executive was less common than that of specialist file clerk. (The position of file executive did exist at the Brooklyn Edison Company, however, as shown in chapter 4.)

Mae Sawyer, director of the Library Bureau's filing schools and uncredited author of *Filing as a Profession for Women,* outlined the difference between a file clerk and a file executive in the booklet's "Self-Analysis Chart for File Executive" and "Test for Position of File Clerk." The two positions shared a need for memory and concentration, but a successful file clerk also had to have good handwriting and the ability to work quickly; a timed test could determine if a candidate had the required dexterity. In contrast, dexterity was absent from the attributes required of a file executive. Instead, the self-analysis chart included lists of physical qualities, mental characteristics, educational assets, and character and personal habits.

Desired mental characteristics included initiative, ambition, resourcefulness, and imagination. The character habits listed included discretion, secretiveness, and loyalty—all intended to emphasize that the position of file executive was one of responsibility.[70]

Moving the focus away from dexterity and file clerks, Sawyer embraced the assumption that "the average woman's instinct for order gives her a peculiar advantage in the field of filing."[71] In *Filing as a Profession for Women* this statement followed a quote from an author named Vance Thompson, who declared, "Woman invented order."[72] This statement appeared in his 1917 book *Woman* (Thompson's previous works were *Eat and Grow Thin* and *The Ego Book*). In *Woman*, Thompson buried his statement about women and order toward the bottom of a page in the middle of a chapter titled "Women and the Sword." Sawyer, however, gave it much greater prominence. She began a chapter in the booklet by pairing the quote with a sentence from earlier in Thompson's book in which he noted that woman's "first achievement was to get man to do things with some degree of regularity."[73]

Unlike a woman's dexterity, however, this perceived natural affinity for order needed to be intentionally shaped and perfected to become the basis of a profession. To be more than a file clerk, a woman had to be curious; she had to take pride in filing and seek to extend it beyond the mere action of putting papers in order. Like all professionals, the booklet asserted, file executives "study constantly the problems that arise in their work, in its many new phases and methods of improvement. They think about their work, and it is the thinking and planning that add interest to any job. If you find work monotonous it is because you are not interested in it; you do not think about its possibilities for growth; you do not try to improve it. It is not the nature of the work that makes it interesting, but the enthusiasm and thought which you bring to it."[74] From this perspective, the profession of filing was no more repetitive than the work of "the physician who has to treat sick people every day" or "the actress in a successful play, where she has to repeat precisely the same lines . . . every day"; resorting to this claim underscored the uphill battle faced by those who were working to get filing recognized as a profession.[75]

The people who actively sought to gain the status of profession for filing founded periodicals and created associations through which they positioned indexing as a science; they wanted the construction of index and classification systems in offices to be the work of file executives. Critical to this project was the magazine *Filing,* which began publication in 1918 and appeared ten times a year for four years, until it abruptly stopped publication in 1922. The two male editors presented the magazine as a "broker" for "information" on filing and a precursor to the formation of a proposed national filing association. The latter did not happen, but from 1920 *Filing* included a column that reported the activities of filing associations that had formed in a number of cities, including Philadelphia, New York, Washington, D.C., Boston, Chicago, Detroit, and Cleveland. Although membership in most individual associations struggled to reach one hundred, the associations existed at least into the 1940s, continuing to hold regional conferences. In the 1930s, with ninety members, the New York City association continued to publish *The File,* a ten-to-twelve-page monthly "bulletin" it had started soon after the demise of *Filing.*

Filing associations typically held monthly meetings (pausing for the summer) that featured one or two invited speakers. Most of the speakers were women who discussed filing practices at their places of work, which varied in type—hospital, magazine, bank, and so on. Common problems also provided topics.[76] In 1921, the Chicago association reported a "lively discussion" on transfer files, made so by the participation of twenty of the sixty people in attendance.[77] Other issues covered at association meetings related to the quality of work. "Personality" was a recurring topic across associations in the 1920s and 1930s.[78] A favorite speaker at the New York Filing Association delivered a talk titled "I Like My Job" in 1921, and in a speech the following year, she stated that "personality, brain power, and interest are assets to success, when combined with observation."[79] Men did sometimes address association meetings, but usually only as representatives of filing equipment companies. In this capacity, they took the opportunity to give sales talks, sometimes demonstrating products. Perhaps befitting its status as an alumnae association (of the Boston School of Filing), the Boston group also provided entertainment at its meetings, usually in the form of recitals or short plays. Titles of the latter included "Pat's Matrimonial Venture" and "Two Clerks—The Right Way and the Wrong Way."[80] By the 1930s, most associations began to broaden the topics addressed at meetings to include such things as fashion and home decorating advice; in some instances, members were invited to recount their international travel.[81]

Despite their efforts, the filing associations failed to displace the dominant idea that filing involved nothing more than putting papers away and retrieving them on demand. Their contention that filing was a profession depended on the claim that the work of filing executives was important to offices and businesses at large. To be valuable, the work of a file executive had to produce value; it could not simply be a service. The value of a filing department was promoted, unsuccessfully, through two arguments. The first was that reducing the time to retrieve files saved money. The file executive was important to this argument because she would have created the system that allowed for the rapid retrieval of files. The second argument invoked the necessity of planning to increase the productivity of a business enterprise (a common argument in filing literature). In an efficient business, any decision should be based on information. Therefore, a business should "rake the entire available field of information to cull ideas applicable to [it]. And that is what a filing system is created for."[82]

These arguments sought to establish a clear distinction between a file executive and a file clerk: a file executive created information, while a file clerk merely retrieved it. However, a major problem for the proponents of the filing profession was that their arguments did not create a clear distinction between file executives and business librarians. For filing to be a profession, it had to encompass all information work (not labor) in an office. However, the business or company library created a space for the very information work claimed by file executives and located it outside a filing department. Commonly found in large enterprises in the banking, insurance, and telephone sectors, such internal libraries existed to gather and store the information their companies needed for planning. Business librarians summarized, extracted, and indexed the contents of documents and books to provide the required information.

Company and business librarians were part of the Special Library Association, which formed in 1909. While some early leaders were unhappy with the association's name, it did succeed in signaling that the members did not work in traditional libraries. Special librarians were different in how they defined their object: information, not books. Robert Williams argues that "providing the information, not just the sources," was fundamental to the professional identity that special librarians created in the early twentieth century.[83] As one early business librarian put it, "For business purposes we tend to dissociate information from literature; we do not want books, we want information."[84] While some special librarians noted the contemporaneous work of Paul Otlet and the European documentalists in explaining their work, others acknowledged Otlet indirectly by identifying their object as "fragmentary bits of information"; this is the approach to information and library work that appeared in public libraries with the creation of "vertical files" (see chapter 4).[85] The editor of a symposium on business libraries that appeared in *Filing* in 1920 used the articulation of information and paper as discrete units in an attempt to chart the commonalities between file departments and business libraries, writing, "The book of tomorrow is made from the pamphlets of today, which in turn, were compiled from the correspondence of yesterday."[86]

While special librarians provided a hurdle to the professionalization of file clerks, they experienced their own professional marginalization. Although the SLA was formed under the umbrella of the American Library Association, there was little cooperation between the two groups by the 1920s. As noted, the work of special librarians was at odds with the book-focused work of the ALA, especially as the association dealt with the growth of public libraries. This division was further accentuated by the fact that, although the SLA claimed the word *library,* in its first twenty years none of its leadership had formal library training.[87] In 1938, the founding of the American Documentation Institute created another professional group with overlapping interests. However, as it developed, the ADI became concerned with broader issues of documentation, a shift that largely complemented SLA members' increasingly more specific focus on the information needs of their particular organizations.

The development of the SLA and the beginning of "information management" left no space for filing to be a profession, despite the middle-class origins of the professionalization movement. Women working in filing departments had limited opportunity to create information. Subject filing offered the best opportunity to assert "brain work" over the manual dexterity of filing. It required some understanding of the content of documents and the workings of the related business, but not at the level of comprehension associated with abstracting or research work. As such, subject filing could be part of a well-run system, but its function was still limited to making information "live" so a filing department would be more than "a depository for old transactions."[88] Company librarians could argue that their work involved "gathering, digesting, and making available . . . the technical and commercial literature of their respective fields," but someone claiming the status of file executive usually had responsibility only for overseeing the work of "making available."[89] Therefore, despite the rhetorical efforts of Mae Sawyer and others, these efforts did not make filing an "active branch of production" and, by extension, a professional track for women.[90]

Filing as information labor was understood as an action, as something akin to machine work. This was a specific kind of machine work that most contemporaries believed was better suited to the hands of women. Thus, employers believed it was not appropriate work for middle-class women; rather, it was work for the hands of women who needed to be taught decorum and norms increasingly labeled middle-class. While a file clerk was expected to be a white woman, her sometimes dubious whiteness made office work a site for broader social and cultural training as gender intersected not only with class, race, sexuality, and ethnicity but also with age.

The articulation of filing as information labor was an encounter with an instrumental conception of information as something particular. It did not preclude the possibility that women who filed could produce information (as did business librarians) or possess broad knowledge about the business. A clerk might have to make quasi-executive decisions that were not explicitly licensed but were nonetheless de facto sanctioned; presentations at filing association meetings suggested as much. The responsibility associated with this kind of work fits within the middle-class pretensions of advice literature, which elevated the private secretary to the top of the clerical hierarchy. However, clerical work could not be the basis for a profession. It was work centered on care, which kept it within the purview of women's domestic work. The work of a private secretary was care directed to the man she worked for. The work of a file executive was often described as "care of the files." In both cases it was easy for a woman to become the "domestic manager in the office," a worker who submerged herself into an all-encompassing pleasantness, accommodation, and attention to detail.[91] The representation of the office or corporation as a "family" further highlighted the connection between office work and women's "essential domesticity."[92] The idea that corporate workers were "family" underlined the belief that in the office, as at home, women were there to assist men, to ensure that men could take on the responsibility their white-collar work demanded.

Figure 6.7. Illustrations from the Library Bureau's 1919 vertical filing catalog helped to establish filing as women's work. Author collection.

Chapter 7

DOMESTIC STORAGE
Cabinet Logic in the Home

The employment of women in offices in large numbers blurred the boundaries between home and office. In both the office and the home, women worked to maintain the spaces to allow men to do what they needed to do. This relied on an understanding of the role of a wife in establishing decorum in the domestic space; she could become a "domestic manager" in the office. While it is unclear if women thought of themselves as domestic managers in the home, it became difficult for women to avoid the values and concerns of the office in their domestic work when ideas of efficiency and productivity started to shape the spatial organization of the home, in particular its storage practices.

In the 1920s and 1930s, vertical filing cabinets were being used in homes, but not in very large numbers. During this period, however, the cabinet logic of designated places and partitions became critical to how objects were stored and rooms were managed in homes. Closets and kitchen cabinets introduced planned storage into the home. The organization of storage spaces in the cause of retrieval affected everything from sugar and spices to shoes and suits, with advertisements and home advice literature explicitly acknowledging a debt to office equipment. As in the office, time became an important value in the home, factoring into assessments of the importance of the workers and the tasks they performed. However, because care was integral to a wife and mother's work in the home, that work came with more responsibility than women's work in the office.

Planned storage contributed to the attempt to give respectability to housework, which included the adoption of another form of Taylorism, so-called scientific housekeeping. Advertisements and exuberant prose presented kitchen cabinets and closets as machines that thought and remembered; like filing cabinets, they were "automatic." Kitchen cabinets and closets did the type of work that servants might have done (and still did in some households). Home advice literature presented the woman who operated closets and cabinets as someone who recognized the importance of managing time in the household. Like the efficiency desk that allowed an executive to open a file drawer without interrupting his phone call, this furniture prevented distraction; it allowed someone to focus on important work, work that derived from responsibility. As a result, kitchen cabinets and closets devalued the work directly associated with the technology. Efficiency in the home, like efficiency in the office, depended on a delegation of labor that relied on gender, race, class, and age norms to create a hierarchy; if a "domestic manager" existed, she existed in a middle-class home.

An analysis of kitchen cabinets and closets illustrates that storage is not neutral. Through the logic of granular certainty storage in the home became the division of enclosed space into smaller spaces. This was efficiency understood in relation to furniture, in relation to technologies and spaces designed to facilitate the movement of objects and bodies. This understanding of efficiency came from the office. Although planned storage in the home was not directly tied to information, it developed from the instrumental investment in the particular that constituted information as a discrete unit. To experience "information" is to experience or perform efficiency as a structure that subordinates parts (the specific) to the whole. Efficiency locates the discrete not in a state of isolation but as part of a system. In the context of paper-based information, the goal was to create a system that would facilitate retrieval on the scale that proponents believed modern life demanded. Applied to the home, this meant that the intentional and systematic storage of cookware and linens put the home in a "state of preparedness," as one writer put it in 1930.[1]

Filing Cabinets in the Home

In 1940, Paul Jerman, an architect and self-identified husband, wrote an article for *American Home* in which he called for the "mass migration" of the "steel file cabinet" from the office into the home.[2] Addressing the magazine's desired upper-middle-class readership, he described a home in which papers lodged in a variety of places could not be easily found. These included Christmas card lists, bills, letters, dog license receipts, recipes, magazine clippings, income tax records, and birth certificates. For Jerman, the difficulty in finding these papers was an information problem; to locate papers properly was to have "information at your fingertips."[3]

The home Jerman described was located in a world that increasingly demanded papers as evidence to verify daily interactions.[4] Paper in the form of documentary evidence and records had become as natural to the twentieth-century home as it was to the office; as Jerman put it, "You have to keep papers, unless you live like the panda in his native habitat without benefit of modern civilization."[5] He was referring to the pervasive demand for information from a range of institutions, paired with the declining evidential value of a person's word in favor of official documents, a shift that occurred in the first half of the twentieth century. This "paperization" of everyday life emerged along with new requirements for passports and driver's licenses, new income tax laws, new banking laws, and universal birth registration.[6] Unlike the middle-class and upper-class travelers who railed against the "passport nuisance" in the 1920s, Jerman was not critiquing the dehumanization of paper or arguing that bureaucracy was causing disenchantment with everyday life.[7] He was not advocating the life of a panda. Rather, he was offering practical solutions for the management of papers so that the home could achieve the standards of an efficient office.

Older paper storage solutions, often in the form of bound books such as family Bibles, cookbooks, and photograph albums, were insufficient for the information demands of twentieth-century society. As Jerman noted, in the nineteenth century life was "simpler"; the "To-no-wah-nee Camp did not insist on knowing exactly when Junior was last inoculated."[8] He believed this problem of paper, information, and identity was an inevitable and necessary part of modern life. It was a mark of civilization's progress, and as such was a problem that could be solved easily through a modern approach. From Jerman's point of view, an individual had a responsibility to manage the different types of paper that allowed access to the benefits of modern society; there was nothing wrong with that trade-off.

As a product of twentieth-century modernity, the filing cabinet provided a solution to the home paper problem because it offered efficiency. As Jerman understood it, the filing cabinet's critical contribution to the creation of system in modern offices was that it managed the storage and circulation of paper through centralization. In the home, it solved paper storage problems because it provided a place for papers that, in its absence, would likely be distributed around the home inside books, desk drawers, shoeboxes, dresser drawers, and vases. "Scattered" was the pejorative descriptor of choice to highlight the problem to which centralization offered the necessary solution. In providing a proper place, the filing cabinet brought integrity to paper storage in the home. Knowing where to find paper saved time; integrity guaranteed retrieval.

Jerman explained the merits of the filing cabinet through the values of the business imagination. He presented the filing cabinet as saving time because it relieved the mind of the task of remembering where papers and other things were. Information, a particular piece of paper, could be found easily through the granular certainty offered by the filing cabinet. Jerman presented the value of the folder as an important information format. A filing cabinet in the home should contain a folder for each member of the family. It should also house folders for magazine clippings on decorating, gardening, and cooking to allow family members, usually wives, to access expertise to help them cultivate techniques for productive living.

It is important to acknowledge that Jerman called his solution a "file cabinet," not a "filing cabinet." This label reflected the shift away from *filing cabinet* in American usage, a shift that deployed the ambiguity of *file* to emphasize the object stored, not just the action. *File* labels both the structure and its contents. In Jerman's article this underscored the content, the file, not the action of putting paper there or retrieving it; rendered invisible, labor disappeared into the function of the technology only to briefly reappear as "fingertips."

However, in his celebration of the "file cabinet," Jerman was careful not to praise it as a machine. It solved the problem of paper in the house, but it was not "automatic." He acknowledged its status

In the Home

The SAFE-CABINET DRAWER-SAFE is the most practical protection for your home valuables.

Legal papers, income tax data, insurance papers, receipts, check and bank books, family budgets—indeed the average home today has more records than business formerly had. And these records vitally affect the savings that years of thrift have accumulated.

Heirlooms, with pecuniary and heart-interest values; letters, photographs and mementos of loved ones; stamp and coin collections; original manuscripts and compiled research, often representing years of study and possibly of tremendous value to the world; silverware and jewelry—All these *warrant* protection.

And, the DRAWER-SAFE gives one hour's CERTI-FIED fire protection to these priceless records and treasures.

You will enjoy having, and rest easier with a DRAWER-SAFE in your home.

SAFE-CABINET
Division of Remington Rand
TONAWANDA, N. Y.

Figure 7.1. This Safe-Cabinet advertisement shows a filing cabinet as part of a home's furniture. The original advertisement is in black and white except for the filing cabinet, which is colored like wood to make it seem more acceptable for the home and the storage of not only paper but also family heirlooms and jewelry. Sperry Rand Corporation, Remington Rand Division records, Subgroup III, Advertising and Sales Promotion Department (Acc. 1825), MSS 1825 III 018 V2 15, Hagley Museum and Library, Wilmington, Delaware.

as a machine or technology only with the comment that there could be an "objection to the file cabinet as a piece of furniture." The solution was to cover it with wallpaper or plywood. He suggested the latter if the file cabinet was part of a "business corner," an area he described as follows: "There, for efficient operation, can be grouped the telephone, writing space, file cabinet, typewriter, with drawers and cupboards for stationery supplies and budget records, and bookshelves for reference books." He continued: "From this one place all the business of the home can be carried out efficiently: ordering, corresponding, telephoning, check writing."[9]

In outlining the business corner and the need to make information immediately accessible, available at people's "fingertips," Jerman offered a new everyday experience of information. While his "mass migration" referred directly to the movement of the filing cabinet from the office to the home, it also highlighted how an informational experience could move from one place to another. The emphasis on the particular, on something that is not knowledge, presented an instrumental encounter with information.

As Jerman's article reminds us, storage is not neutral. In making objects readily available, storage technologies create specific relationships to those objects. The arrival of a filing cabinet in the home contributed to the promotion of efficiency as a way to live successfully in "modern civilization." The articulation of information and efficiency often centers on a distinct understanding of storage and memory. To consider the relationship between storage and memory is to isolate what is particular to cabinet logic in this period, what is distinctive about *modern* storage. Therefore, a comparison with the older forms of paper storage, declared inadequate by Jerman, highlights the larger implications of the migration of this new mode of information storage to the familial space of the home. Older forms of storage within the home offered a different relationship to memory, a relationship not mediated by efficiency and productivity.

The Family Bible in the Nineteenth-Century Home
Bibles have had a presence in American homes since the nation's founding, but it was in the period after the Civil War that the so-called family Bible became "a repository for family memories."[10] The family Bible arrived as part of a general trend toward domestic sentimentality and affectionate religion in the home. Beyond facili-

tating the recall of religious lessons, the Bible became part of the articulation of domesticity, piety, social propriety, wealth, learning, and refined sensuality that Colleen McDannell calls "material Protestantism."[11] The family Bible's prominent location in the parlor, along with its size (many were as large as 10 by 15 inches and weighed ten to fifteen pounds), made it an object of display, a function often enhanced by leather bindings and gold clasps. To enable appropriate presentation, publishers and church groups circulated instructions for building parlor lecterns, installing marble and wood brackets, and embroidering tablecloths to be placed under Bibles.[12]

The size and integrity of the family Bible also facilitated a secondary role for it as a depository for family memories, following a tradition of using bound books to record dates and store loose paper.[13] The family Bible's incorporation into this history began with the practice of writing the birthdates of family members on blank pages. Printers responded by adding pages to Bibles specifically for the recording of family events such as births, marriages, and deaths.[14] From the middle of the nineteenth century, publishers added more illustrated pages to be used to record actions and activities such as temperance pledges and membership in fraternal societies. These Bibles also became secure places for families to store loose objects that might otherwise be misplaced, items of sentimental value (poetry, hair clippings) as well as practical records and other papers (bills, receipts, insurance policies).[15]

Although the family Bible functioned as an instrument of family record, this did not disrupt its central role as an anchor for the family's spiritual life. Therefore, placing objects in a Bible not only preserved them but also made them sacred. A Bible recorded milestones as important events in the family's religious life; it connected the life of the family to a greater purpose. It was not intended to turn family memories into an efficient encounter with "information." Nonetheless, some publishers included tables in Bibles expressly for the recording of important events. Systematizing the practice of recording births, deaths, and marriages in this way resulted in documentation that, in the absence of official records, could potentially be useful to governing institutions. For example, prior to the establishment of universal birth registration in 1930, the U.S. State Department occasionally accepted handwritten notations in family Bibles as evidence of date and place of birth for passport applications.[16]

In the early decades of the twentieth century, the family Bible became less significant as a religious object and as a storage container. As McDannell notes, domestic ideology moved away from a focus on faith, contemplation, and limited material possessions. The home became a more casual and informal space, where rooms were rethought to serve multiple purposes. The "living room" replaced the "parlor," along with its role as a space for serious contemplation and interaction. The living room was a rationally planned space open to technology, a space that, according to lifestyle experts, allowed people to live efficiently in modern society.[17]

These changes challenged the function of the family Bible as a place of storage for memories. As Jerman suggested, the growing number of important papers demanded specialized storage; the Bible was no longer adequate. Recognizing the loss of this role, Bible publishers reduced the space for family records to simple lined pages for noting births, marriages, and deaths. In addition, as portable cameras became more affordable, the photograph album offered a new place to store memories, although it did not offer a new type of remembering. Rather, it extended (and made explicit) the family Bible's mode of memory: the photograph album functioned as a prompt for family members to tell stories to explain its contents.[18]

This embodied form of showing and telling separated the photograph album and the family Bible from how "information" was experienced in the domestic and commercial filing cabinet. These different modes of memory created distinct experiences of knowledge and information. As Martha Langford observes, the photograph album requires a personalized mode of remembering that depends on a "pattern of knowledge and experience inextricably linked with the self."[19] In contrast, the filing cabinet's perceived value is that a "stranger to the file" can access its content. The content of a filing cabinet, as information, is understood through its location in a system made visible in a cabinet drawer, not through personal memory. This distinguishes "associative" human memory from the automatic memory of technology that "recalls" through an address. The latter delegates aspects of memory to a machine; it relies on information to make access no longer dependent on the need to know content. Understood as a change in modes of memory, the movement of this experience of information from the office into the home via changes in home design and furniture produced an alternative to "sentimental memory"; Lynn Spigel identifies the significance of this by arguing that it was a move "from memory to storage, from Victorian sentiment to rational modernism."[20]

Planned Storage

In the middle of the twentieth century, industrial designer George Nelson called planned storage in the home "active storage," echoing office equipment literature from earlier in the century.[21] Planned storage entangled the domestic and the commercial in a new way to produce spaces in the home that organized aspects of sentimental memory (closeness, understanding, and detailed knowledge) through standardization, procedure, and efficiency. Or, put another way, the practices and functions of modern storage produced a logistical memory that existed alongside sentimental memory. Logistical memory is about remembering where something is. This can be achieved through the delegation of that task to specific structures, such as cabinets, compartments, and specifically shaped shelves. Memory becomes a problem defined and solved through granular certainty; it is not centered on narrative and remembrance.

Of course, the use of furniture and spaces devoted to storage in the home did not begin in the early twentieth century, when closets and cabinets designed to store specific objects were factored into the design of houses. Prior to that period, however, a different set of concerns shaped storage problems. In the nineteenth century, portable storage began to expand beyond the homes of the wealthy, as dressers (which replaced chests by adding drawers) and other furniture intended for storage appeared. Designated storage areas in most homes were leftover spaces, usually around edges of rooms and above and below stairs. In smaller homes through the nineteenth century, open shelves were important sites of storage.[22]

With limited storage technologies, nineteenth-century domestic advice books advocated ideas of system to organize work in the home. As Melissa Gregg argues, the appearance of these techniques prior to their adoption by Frederick Taylor and his disciples troubles assumptions that the first managers were male engineers.

Mid-nineteenth-century domestic advice authors deployed ideas of system and time management in an attempt to connect women's domestic work to nation and religion; religion in particular provided the sense of purpose used to prioritize tasks in the home.[23] Gregg shows that the success of many of the time management strategies proposed by these authors depended on the ability of their well-off readers to delegate tasks to other women.[24] By the 1920s, however, this delegation was increasingly directed to new household storage technologies rather than to other people. As the filing cabinet illustrates, this shift meant that any remaining labor associated with specific tasks was (further) devalued or made invisible to allow the new technology to be represented as efficient.

Therefore, new designs for closets and kitchen cabinets taught their predominantly female users to value efficiency, which led to changes in ideas of self and productivity. The shift in focus from "religious devotion and duty directed to the needs of others to ideas of efficiency based on self-improvement and individual accomplishment" occurred in the same cultural context that caused a functional change in family Bibles and "personality" to replace "character."[25] As a result, Gregg contends, efficiency became a way to comprehend moral perfectionism, where "perfection" was understood as doing as much as possible in the least amount of time.[26] Addressed in this way, these women were part of "a necessary reconfiguration of attitudes toward self and work performance" that enabled the "gradual adjustment to productivity as common sense," an adjustment that Gregg argues gives productivity its twentieth-century history.[27] Domestic space was to be arranged according to how productivity defined storage needs. Compartments and partitions became the organizing principle of storage; cabinet logic deployed in the name of efficiency prioritized logistical memory.

Closets

Until the end of the nineteenth century, closets were an afterthought in home design, little more than wall cavities that tended to be deeper than they were wide or wider than they were deep, making storage and retrieval difficult.[28] Prior to the middle of the century, clothes were typically hung from pegs or stored in chests, or sometimes hung in wardrobes and armoires, depending on a household's social status.[29] The "closet problem" emerged at the turn of the twentieth century, with a sustained critique of the existing scale and nature of storage in the home: there were not enough closets, and those that existed were not where they were most needed, and it was difficult to get things in and out of them.[30]

As a storage technology, a closet functions to put storage beyond the edge of a room. It is often hidden with discreet doors that accentuate a tension between visual concealment and physical access.[31] The closets that arrived in early twentieth-century homes sought to conceal the increasing numbers of things people purchased as many Americans became consumers of mass-produced goods. The availability of affordable ready-to-wear clothes and changing leisure patterns meant that people had more items of clothing as well as other goods that needed storage. However, while the closet concealed its contents behind doors, its status as a problem had more to do with the interior. In the early twentieth century, the women who wrote about the "closet problem" in middle-class magazines focused on the "proportional relationship between the closet and the goods to be stored."[32] People wanted closets to be efficient in both space and time. The problem was access, and compartmentalization was the solution.

Retail display and the office provided examples of the benefits to be gained from giving different objects their own specifically shaped storage spaces. Both focused on grouping similar objects together, with the office providing inspiration for the concealment of objects rather than their display.[33] Writing somewhat ahead of his time, an 1890 closet critic promoted the design of an office desk as a model for home storage, celebrating the fact that "there is nothing which may come to it which has not been considered. There are provided many receptacles, drawers of all sizes and kinds, properly divided. Closets should be considered in the same way."[34] In contrast, he argued, existing closets were, at their best, "little more than empty zones of space with a few hooks around their edges and shelves for items that could not be hung." In denigrating the closet, he was touting the benefits of something like the Wooton Desk, which was about to be replaced by the horizontal filing cabinet, with its different-sized drawers to house the range of documents used in an office.

In the early twentieth century closets changed according to the belief that successful home storage required the rationalization of space. This could take the form of a closet designed to store one kind of thing or a closet designed to store multiple things, each in its own specifically designed space. The key was an interior space that anticipated the objects to be stored, so they could be deposited and retrieved easily, according to the increased tempo that modern life demanded. The architectural definition of a closet as "anything which encloses or includes" captured the purpose-built spaces that appeared in homes for things like ironing boards, sewing machines, and umbrellas.[35] These closets hid things considered ugly (sewing machines and ironing boards) when they were not in use, as well as things that simply got in the way and affected the efficient movement of people through the home. Following the mantra of "never crowd the hall," a *House Beautiful* article asserted that the ideal home needed designated (and designed) spaces to get umbrellas, hats, and coats out of sight (one suggestion was an umbrella closet in the square post at the bottom of a staircase).[36]

Closets for clothing were designed to give different items their proper places. For one champion, a modern closet needed "boxes for hats, with standards procured from the milliner. . . . Long drawers for skirts, short drawers for waists, and boxes or pigeonholes for shoes,—in short a place for everything."[37] Ideally, the long drawers would be long enough to lay out "dress-skirts" without folding them—as in the office, folding had come to be seen as inefficient, something to be avoided in order to improve ease of access and protect objects from damage. A similar rearrangement of space occurred in linen closets, which increasingly partitioned space for different-sized objects, with access further improved via sliding shelves or flap shelves.

When closet manufacturers targeted male consumers, the influence of the office and granular certainty could be even more explicit. In the 1920s, a British company marketed its closet system with the imperative "Gentlemen file your apparel." The advertisement continued: "The old spike file has [no] place in the equipment of the modern man of business. Its day is past and the vertical filing cabinet has become a necessity in the well-ordered office. The same principles of orderliness are now being applied to clothes." Compactom, another British-based company, adhered even more closely to the logic of the filing cabinet. Its patented system involved placing folded shirts in large file envelopes, which were then stored standing "vertically" on shelves in a closet to "allow for the removal of one or more without disturbing the balance."[38]

Kitchen Cabinets

By the 1920s, "experts" began paying as much attention to kitchen cabinets as to closets, but they more explicitly linked the former to the ideas of scientific management. An awareness of movement in the kitchen contributed to new ideas for the more systematic location of food and equipment. In practice, this often meant ideas derived from motion studies, and less frequently studies of the steps women took while working in kitchens. Articles in homemaking magazines were populated with phrases like "routing of work" and "assembly-line principle." A *Better Homes and Gardens* article title emphasized that it was important to organize "kitchens in logical order." Or, as a title in *Ladies' Home Journal* put it even more simply, kitchen design was about "better places to put things." Kitchen design had become a thing.[39]

Kitchen design meant arranging kitchens according to "centers" that brought together equipment, ingredients, and specific tasks such as preparation, cooking, dishwashing, and serving. Although applied less frequently, the label "departments" made business and industry's influence on the kitchen even more apparent. In the words of a *House Beautiful* article, the need for centralization stemmed from a belief in "the elimination of fatigue through having things grouped."[40] The kitchen user's fatigue could result from an excessive number of steps or the frustration of searching through poorly designed cupboards, drawers, and pantries.

Supplies and dishes were stored to create a center for food preparation. In choosing a kitchen cabinet as a center, homeowners were advised to assess the convenience of the interior arrangement. As cabinets developed, kitchen storage became the strategic use of shelves, drawers, and partitions. As in the office, the emphasis on space and movement provided the conceptual inspiration for new storage furniture that prioritized retrieval through visible divisions that designated places for objects. In an echo of how office equipment companies sold filing cabinets, one manufacturer described its cabinet as "a kitchen with brains"; the late nineteenth-century articulation of efficiency and cabinet logic had arrived in the kitchen to give a cabinet automatic memory.[41]

Figure 7.2. This 1922 Kitchen Maid advertisement is an example of the kinds of marketing appeals that appeared when kitchen design began to be influenced by "home science experts" in the 1920s. Kitchens and the work performed in them became subject to the same concerns about productivity that characterized scientific management in factories and offices. Author collection.

For experts of modern kitchen efficiency, older drawers were too deep for objects in them to be found easily. One common strategy was to divide drawers into sections to group related items together. Shelves in cabinets and cupboards were made removable, stepped, or sliding to make it easier, or unnecessary, for users to reach inside the cabinets. Acknowledging the lessons learned in the office, one home help article offered the example of removable partitions in a deep drawer that "made it into a file box." Vertical or upright partitions added to shelves allowed plates and platters to be stored on edge: "You can pick out with one hand just the dish you want!"[42] Not surprisingly, these were called "files." "Sliding partitions" for storing kitchen linens copied the trays used to store blueprints in architectural offices.[43] Similar attention was paid to the interiors of bathroom cabinets, which developed at the same time. Deanna Day has suggested that these specifically designed containers encouraged the careful curation of objects as well as the belief that a house should be well stocked. They gathered different objects (e.g., thermometers, razors, and toothbrushes) to make them of one kind. Cabinets underneath sinks, as well as closets, were used for other things that were considered more private, such as menstrual products.[44]

Hoosier Cabinets

The Hoosier cabinet represented the apotheosis of the early twentieth-century fascination with order in the kitchen. Because almost all wooden kitchen cabinets of this style were manufactured in Indiana, it took its name from the nickname for the state's residents.[45] The design of these cabinets—meant to provide sufficient storage space in the 120 square feet typically available in the kitchen of a bungalow or cottage—drew directly from office equipment and furniture.[46] A Hoosier cabinet was usually 72 inches high, 48 inches wide, and 22 inches deep. In appearance, it was similar to a popular style of horizontal filing cabinet. The horizontal filing cabinet, which included drawers sized for different types of paper used in business, provided a useful model of compartmentalization. However, "verticality" was vital to the Hoosier. Even more than a horizontal filing cabinet, it utilized height to subordinate parts into a coherent system.

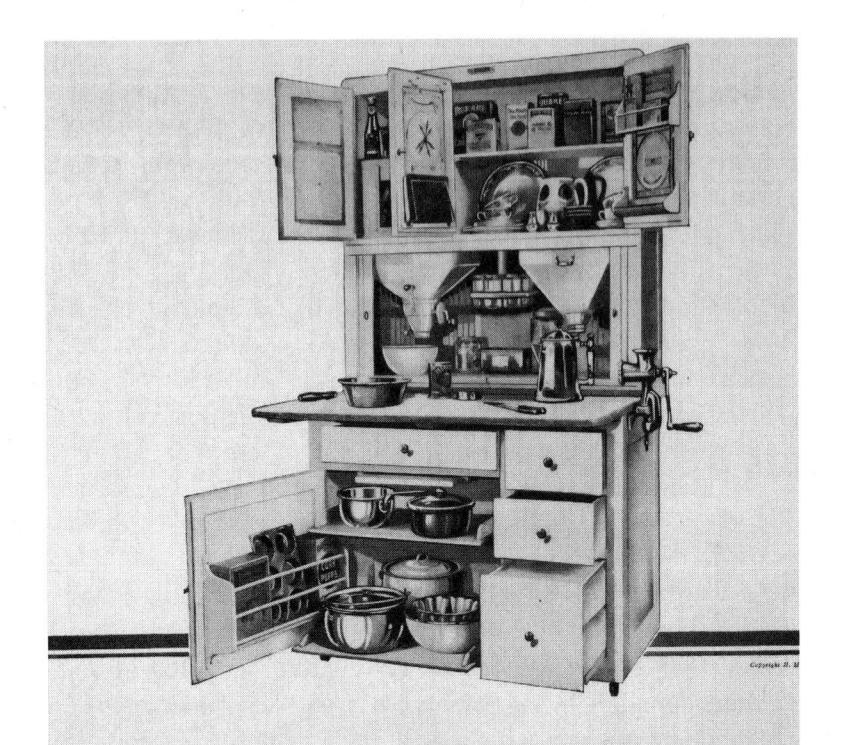

Saves miles of steps

NEARLY two million women use the Hoosier every day. These careful housekeepers save many steps each day. And when *you* are ready—and you will be sometime—the Hoosier will save work for you.

Every type of bin or food container has been tested by Hoosier. Those that were practical have been adopted—the others discarded.

But to realize how much time the Hoosier will save, you must see it. Go to a Hoosier store and sit in front of the Hoosier you like best. See for yourself how much more pleasant your work can be made.

But don't delay. If you are not sure of the name of the dealer who handles the Hoosier, write us. We will direct you to the proper store.

HOOSIER

THE HOOSIER MANUFACTURING COMPANY, 220 Jackson Street, New Castle, Indiana
The Hoosier Store, Mezzanine Floor, Pacific Bldg., San Francisco, California
The Hoosier Store, 368 Portage Ave., Winnipeg, Canada

Figure 7.3. This 1922 advertisement for a Hoosier kitchen cabinet shows how it was designed to hold everything the company believed a woman would need in a kitchen. The advertisement's tagline, emphasizing the saving of time, underscores the influence of scientific management on the design and marketing of kitchen cabinets. Author collection.

Accommodating a range of different-sized objects, the Hoosier brought together all the different techniques that characterized the modern kitchen. Mary Anne Beecher uses the concepts of "consolidation" (grouping related items together) and "disclosure" (making the locations of tools and supplies easily known) to illustrate the connection between the new storage regime of the kitchen and the earlier transformation of office furniture into office equipment.[47] In the argument of this book, *consolidation* and *disclosure* refer to efficiency and its faith in granular certainty. As Beecher describes the Hoosier cabinet, large shelves and drawers housed such items as pans, kettles, and nests of mixing bowls. Smaller divided drawers held cutlery, kitchen linens, and packaged goods. Special metal boxes stored breads and cakes. Tilting or sliding dustproof bins with dispensers held sugar and flour. Little glass jars for spices revolved on a rack, and hooks and tiny shelves for additional small containers and kitchen tools lined the doors. Promotional materials claimed that a Hoosier cabinet had "four hundred articles all within arm's reach."[48] According to Beecher, when the Hoosier cabinet dominated sales, most kitchen cabinet manufacturers sought at different times to associate their products directly with business or manufacturing environments.[49] In the world of the kitchen cabinet, to be "businesslike" was to be efficient. In their advertisements, manufacturers used the office to emphasize their laborsaving claims without noting the popular understanding of clerical work as monotonous.

An advertisement for Seller's stand-alone cabinet used the tagline "The business of getting meals." An advertisement for the Hoosier-style Napanee Dutch Kitchenet paired a small cutout of an office with the claim that the cabinet "makes kitchens businesslike." The small, almost intimate, office was attractively laid out with a couple of desks and a filing cabinet; a skyline of skyscrapers was visible through a tall window. However, the main text of the advertisement did not directly mention offices or office work. Rather, it developed the claim that the cabinet "makes every motion count," noting that "the superiority of the Napanee Dutch Kitchenet was attained by what efficiency engineers call scientific 'motion study.'" The advertisement elided any association of motion studies with repetitive manual labor. By implication, it articulated motion studies as a mode of efficiency and organization that would help the worker.

Figure 7.4. This 1921 advertisement for Napanee kitchen cabinets from *Good Housekeeping* shows the impact of office equipment on kitchen design. Most kitchen cabinet manufacturers sought to link their products to filing cabinets and the efficiency they believed people associated with modern offices. Public domain/Google Books.

Other advertisements made this connection more explicit. In line with attempts to show that stand-alone kitchen cabinets prevented fatigue, the Hoosier Company's annual $250,000 advertising budget put particular emphasis on the claim that using one of its cabinets produced an "equable and cheerful temper and tone in the housekeeper." The claim was that, as in the office, a partitioned cabinet would do work that a person once did.[50]

The delegation of logistical memory to a cabinet through consolidation and disclosure was not limited to the two million Hoosier cabinets sold in the 1920s. These sales indicate that Hoosier kitchen cabinets made it into only 10 percent of American homes, but the storage techniques associated with them shaped kitchen storage solutions in this period. Those techniques were critical to the interiors of the built-in cabinets that began to be installed around the perimeters of kitchens in the 1930s. A Cornell University report of homes in upstate New York from that decade found that even if people were not buying kitchen cabinets or built-ins, they were interested in altering their existing furniture to follow the granular logic that shaped the modern kitchen.[51]

Home as Office

The positive parallels drawn between housework and office work also contributed to the call for the busy housewife to have a home office. Moving away from the language of efficiency and business as a justification, one writer asserted that a wife and mother "does need 'A Room of One's Own' as Virginia Woolf so conclusively proves."[52] While Woolf may or may not have provided conclusive proof, in practice this "office" was demarcated not by walls but by furniture in the corner of a room (as Jerman would later suggest). In domestic experts' ideal kitchen, a desk constituted the "planning center." But the kitchen desk, while more common after World War II, never became widely popular.[53]

The call for desks (and filing cabinets) in the home was part of an attempt to change the status of housework. The proposed desk was that of a manager, not a clerk. It was intended to identify a woman's housework as "professional labor" and, as Dianne Harris perceptively notes, "distinctly white labor."[54] A desk with paperwork distinguished the woman of the house from other women, usually "lower-class immigrant workers, who might be hired to perform menial labor."[55] This attempt to conjure a white-collar atmosphere obscured the reality of the tension between the roles of manager and clerk as they manifested in homes. Most women had little or no hired help, and their housekeeping labors were more akin to the manual work of clerks than to the work of managers.

Scientific Housekeeping

Scientific housekeeping literature intentionally associated office work and domestic work in an attempt to raise the status of domestic work. These writings return us to the explicit experience of information encapsulated in the filing cabinet and card index. Christine Frederick, the most famous promoter of scientific housekeeping, asserted that the home, like factories and offices, could not run smoothly without "immediate and reliable records"; she borrowed this phrase from a Frederick Taylor disciple, Harrington Emerson, who listed it as one of the chief principles of efficiency.[56]

Following Emerson, Frederick's writings exhibited immense faith in the recording, storage, and retrieval capacities of the index card. In Frederick's hands index cards became a "time-and-worry-saving household file." She suggested that readers would be surprised "how much information is packed away into that little drawer fifteen inches long."[57] Frederick described a system based on eight main categories that could be broken down into numerous subcategories: accounts, addresses, medical, house, library, financial, house hints, and general inventory.

Frederick openly acknowledged that the organizational structure of her card system came from her admiration for the way businessmen "preserve their data and information."[58] In explaining the usefulness of creating information on index cards, she lifted verbatim the ideas from office management literature and office equipment advertising copy that have appeared throughout this book, and with them the privileging of an instrumental encounter with a discrete unit she called "information." Adhering to the logic of the office, her cabinet and its supplies demanded uniform records (getting rid of separate books and papers). Therefore, a "card file" centralized records in one place, making information easily accessible. It also allowed for the (inevitable) growth of records as a home became more efficient and productive.[59]

In her magazine columns and book chapters on record keeping in the home, Frederick offered examples intended to make clear how delegating memory to a card produced information that could be easily extracted and circulated. For instance, she proposed that a woman going shopping should take along, clipped together, the cards that recorded each family member's measurements and sizes, in case she saw some "good underpriced socks" but could not remember her husband's shoe size.[60] Frederick also enjoyed recounting the time she rescued a dinner party from boredom by retrieving her quote and poem file. She called this "entertainment by card system."[61]

However, for Frederick the blurring of home and office was best limited to the application of office techniques and technologies to the home. She did not champion women, especially married women, working outside the home. Specifically, Frederick struggled to see how a woman would be satisfied working in an office. Not only did she consider the pay inadequate, but she also believed that office work occurred in "manifestly unaesthetic surroundings." Therefore, a woman would surely prefer to work at home, in "as aesthetic a background as she chooses to make it."[62] However, the "self-expression" that Frederick identified as available to women working at home also included ideas taken from the office: the "self-improvement and accomplishment" that redefined housework in line with efficiency.[63] Frederick explained her strategic use of office management techniques in terms that defined self-interest through time management and an individual's relationship to productivity: "I found finally that I just couldn't afford to waste the time and nervous force usually spent in putting away, hunting for and hauling things out of corners to find just the right thing."[64]

Home, Not Office

In contrast to Frederick, drama critic and author Montrose J. Moses expressed some wariness toward rethinking the home (and domestic storage) through productivity. In a 1930 essay that appeared in *House & Garden* magazine, Moses explicitly offered a "plea for system," but he did not want ideas of system and productivity defining the home, especially in the case of "family memory." To this end, Moses called for a system of built-in "pockets" for storing mementos.

He introduced his ideas about home storage by noting that the suit he was wearing while he was writing had fourteen pockets, and when he was outside his overcoat gave him six more. The multiplicity of pockets created problems, however—to be precise, "twenty havens through which I have to search with panic when I desire to locate a particular memorandum."[65] (Women did not have this problem as their clothing came with few pockets, most of which were decorative, intended for display, not storage.)[66]

The problem of fumbling through a lot of pockets made clear what constituted Moses's "plea" as well as the limitation of his use of pockets as a storage metaphor. Although pockets provide storage space in a suit, a suit is not organized to store objects—the exception being the short-term storage that a suit provides when worn, when it serves as "a complete envelope for the body," as Anne Hollander notes.[67] In contrast, Moses wanted the space of the home to be intentionally organized as a space of storage, with each pocket "a demarcated place for gathering."[68] This was a call for system and centralization against the scourge of scattering.

However, Moses remained vague about what house pockets would look like. In passing, he offered the idea of a pocket for every decade of a person's life. Inspired by banks, he envisioned pockets in the form of drawers in baseboards. Inspired by Prohibition tales, he proposed the idea of storage spaces behind portraits, with the images identifying the owners of the objects stored there.[69] Thus, while he was unclear about the exact design of his pockets, Moses was clear that he wanted them to be present in the house, separated but connected. Pockets would therefore be more accessible than an attic, where families frequently "dumped" mementos, making them more easily available to mice than to people. As part of a system, pockets would offer smaller designated spaces that would make individual objects easier to find than in an attic. Therefore, Moses likened the presence of pockets in the home to the interior organizational structure of mobile forms of storage (largely used by men): "the game bag, the mail pouch, the brief case, the music roll, the medicine chest, the tool kit, the wallet."[70]

Moses believed that a system of pockets would be valuable because it would put a stop to the feeling that life was one long memory test regarding where things were put. The pockets, with their "definite locations" for objects, would ensure that "the logic for putting things away" would not be forgotten. However, Moses was making a "plea for system," not a plea for productivity. He assumed that "the ease of location would . . . enhance the joy of remembering about things."[71] Moses emphatically believed that system saved time; he simply wanted it to be applied to mementos, to be used in the service of sentiment rather than for the efficiency of a "well-ordered present." The "orderly present" was feminine labor, housework. Moses noted that architects ensured that "closets take cognizance of things that are: here is ample space for the stove polish and the floor mop, there is a special shrine for the endless rows of shoes and slippers." Without naming scientific housekeeping, he criticized it in commenting on the hours that had been spent "concocting sliding drawers and shelves for linen." Writing as someone who likely rarely needed to look for linens, he believed such efforts should instead be applied to storing things that mattered to a person's character, like locks of hair, stones collected in childhood, and love letters.[72] The purpose of house pockets was to enable a man to "renew acquaintance with his Buried Self." Significantly, Moses made no mention of the filing cabinet as a solution to the problem of storing love letters. While he advocated "filing" socks, ribbons, and ties in a "sequence of wearing order," he restricted the influence of the office to providing metaphors for the action of storing; the presence of actual office-type storage containers risked turning the home into the office.[73]

The omission of a filing cabinet spoke to Moses's belief that pockets should be both discrete and discreet to ensure that system should not rule a home. In writing that "house pockets should not dominate the atmosphere," Moses expressed wariness about system, efficiency, and productivity providing the value system for everyday life. For him, a "home should be attached to sentiment," not simply to efficiency for the sake of productivity. A filing cabinet was too overt in its connection to efficiency. Therefore, the filing cabinet could not be a model for a pocket, nor should it be a model for the mind. Having made his plea for system to be applied to mementos, Moses concluded his essay by acknowledging the difference between memory and storage: "I know we cannot live by exactness. Memory must be allowed to float." Unlike objects, memories cannot be fixed in a definite place. Moses noted, "You can't expect yourself to say, when you give your wife the first kiss, 'File that, my dear, for future reference!' Much more consistent with human nature is it to say, 'Confound it, where did we put our marriage license, after we went to housekeeping!'"[74]

My purpose in concluding this history of the emergence of the filing cabinet with an analysis of changes in household storage is not simply to highlight the similarities in cabinets and the use of office images and metaphors to sell home storage. My primary aim is to illustrate how, through storage technologies, an early twentieth-century conceptualization of granular certainty (and associated economic values) became pervasive beyond the office. That is, the story of the emergence of the vertical filing cabinet is in part the story of how filing as a mode of labor and organization became an element of everyday life, and what was at stake in this change. In struggling to explain his plea for system, which included the presence of filing but the absence of the filing cabinet, Moses illustrated the important role the information order of the office played in shaping how people made sense of the early twentieth-century articulation of labor, efficiency, and productivity. For Moses, it was okay to treat socks, ribbons, ties, and linens as paper and "file" them in a logical and rational manner. It was also acceptable to organize memories as mementos to ensure the past could be easily remembered. However, the organization of mementos should not be subsumed under the values that Moses associated with filing. His fear that a system based in a narrowly defined, economic understanding of efficiency would destroy the distinctiveness of the domestic space left no place for a filing cabinet. Moses thought it absurd that such things as a kiss might be subject to the articulation of efficiency and productivity that filing represented for him.

Moses's fears about the office becoming entangled in the home via the economic values inherent in modern storage practices were realized in a version of his pockets that entered homes in the middle of the twentieth century. American designer George Nelson's category of "active storage" originated in what he called the Storagewall. This popular midcentury storage technology was a modular, built-in unit designed to store objects associated with the everyday life of the home. Models varied, but among the components they might include were a "games closet; bookshelves and magazine racks; a space to store vases; a wet closet (for 'high balls'); a fold-out desk with cubbies for family records (meaning bills, receipts, and letters); a built-in radio; speakers; and a drawer unit for a record player."[75] Initially designed to occupy the airspace inside walls (pockets, in Moses's imagination), subsequent Storagewalls were built as freestanding cabinets. As a modern storage technology, the Storagewall was designed to target what Nelson viewed as the "clutter" that dominated modern homes (especially smaller suburban homes without attics). He wanted the Storagewall to house everyday items, to ensure those items were available for instant use. Nelson's embrace of efficiency blurred the acts of retrieval and use as he sought to use the Storagewall to shape both acts in terms of speed and circulation—hence the concept of active storage. To store something is to anticipate the future; however, in a modern storage technology like the Storagewall, the future becomes the contemporary, daily life becomes modern. With the timeline of the future shortened, storage space is prioritized for in-demand everyday objects. In contrast, "the sentimental things of everyday life," rarely looked-at keepsakes, were deemed frivolous and had no place in a Storagewall or a home. Nelson turned household pockets into Moses's worst nightmare.[76]

Lynn Spigel insightfully identifies Nelson's investment in active storage and retrieval as an instantiation of "an information logic where things can be filed and retrieved on demand."[77] As I have argued in this book, this information logic emerges out of an investment in the particular. That is, the Storagewall developed from the early twentieth-century faith in granular certainty; because of her investment in a different set of arguments, Spigel connects the Storagewall and information to create a sophisticated prehistory of ubiquitous computing. However, the information logic that Spigel identifies has its own history, which, as I have demonstrated, can be told through the filing cabinet, closets, and kitchen cabinets.

The responses of Moses, Nelson, Frederick, and Jerman to questions of storage in the home illustrate that when planned storage appeared in the home, the focus was on making discrete objects easily available. Although Frederick never directly compared a Hoosier cabinet to her "memory cabinet" of index cards, she championed the former by emphasizing attributes that *System* magazine associated with index cards and the organization of information: "No kitchen can be standardized if there is not a definite place for each article. Nothing must be overcrowded, nothing jostled with other articles."[78] Frederick accepted that it was necessary to apprehend the world in granular terms in order to create the certainty and reliability on which (the increasingly commonsense idea of) productivity was based.

In the first half of the twentieth century, storage became the organization of an enclosed space into smaller pieces of space. Through rationalization, storage was intended to facilitate finding something specific. However, as in any space, storage technologies in the home structured particular relationships with their contents; storage is not neutral. This was a conception of storage that adhered to cabinet logic imagined according to late nineteenth-century ideas of system, productivity, and efficiency. It is critical to acknowledge that this iteration of cabinet logic came out of the office, a space organized primarily through the dominant ideas of capitalism and gender. To emphasize that place of origin is to argue that the discreteness, the particularity, of a place for this pan not that one, sugar not flour, was indebted to a historically specific conception of storage logic; it emerged as a response to the rethinking of information and storage that produced the manila folder and the filing cabinet.

In the home and in the office, the activation of modern storage technologies demanded a set of values that linked identity to productivity and efficiency. The partitions of a Hoosier kitchen cabinet directed a person to particular locations to place or retrieve certain objects. To use a Hoosier cabinet was to experience uniformity and precision, to enact a mode of behavior and a stance described as "efficiency." The speed and efficiency associated with retrieval emphasize that modern storage technologies emerged from the convergence of object and hand, rule and protocol. This is why it made sense to present storage technologies as machines that did the work of organization, of creating a system, so that people did not need to think or remember. These machines required a certain form of labor performed by a particular worker. An encounter with these technologies encouraged people to understand their self-worth through productivity; as hand and object became the representation of efficient labor, these economic values became a way to express identity.

Therefore, although changes in domestic storage did not address information as a formal category, they do offer a suggestive example of how people experienced information logics through the material consequences of modern information. This highlights the role of storage in manifesting the commitment to particularization that shaped the conception and use of information at the turn of the twentieth century. The development of planned storage in the home, the division of larger spaces into specifically shaped smaller spaces to facilitate timely retrieval, linked the particularity of information to efficiency. That is, efficiency shaped this moment in the genealogy of the ascendancy of information. Therefore, to isolate this moment is to construct a material history of efficiency. The reconfiguring of objects and practices according to a gendered conception of efficiency gave information a visibility it did not previously have. An instrumental form of knowledge became "information," a discrete unit available on demand and valued for its efficiency.

Afterword

OUT OF TIME, OUT OF PLACE

In Burlington, Vermont, on a weedy lot owned by the city, stands a stack of eleven metal file cabinets a little more than forty feet high.[1] The cabinets contain thirty-eight drawers, eight of which are partially open. A travel website has named the structure "The World's Tallest File Cabinet." The local architect and gallery owner who erected it in 2002 has another name for it: *File Under So. Co., Waiting for* . . . When she created it, forty-five-year-old Bren Alvarez intended it to satirize and symbolize "the bureaucracy of urban planning." The thirty-eight file drawers represented the thirty-eight years the Southern Connector road project had been under review. Alvarez got the file cabinets from a local business that was giving them away. Some of the cabinets were vintage, with brass nameplates and handles on the outside and springs and levers on the inside. Their weight somewhat negated their charm, especially in contrast to the newer cabinets, which Alvarez described as "light as tin" and much easier to work with.[2] Alvarez welded the file cabinets together and used an interior steel post to stack them in the middle of the path of the proposed roadway; they would have to be removed if the Southern Connector was built. More than eighteen years later they remain in place, with the roadway still under discussion.

The sculpture has become a landmark, seen on local art projects and on T-shirts; hundreds of photographs of it have been posted online. In 2019, a local group organized a ninety-minute "worship"

Figure A.1. Bren Alvarez, *File Under So. Co., Waiting for . . .* (2002). This sculpture in Burlington, Vermont, has also been called "The World's Tallest File Cabinet." Photograph by David Leff.

service at the base of the cabinets during which attendees confessed their "sins of disorganization" and had their foreheads anointed with correction fluid. But that year also saw the Burlington City Council finally approve a design and construction budget for the road project, now called the Champlain Parkway. Although the Burlington mayor declared, "The time for debate, amendment, and appeal has long passed," state and federal agencies delayed the project at the end of the year. In the midst of this, Alvarez, sensing the roadway might finally be built, found a new location for the tower on a friend's property. Wanting to move it before a demolition crew showed up, Alvarez took her idea to the city council office only to discover that because of the height of the sculpture she had to apply for zoning variances. As of this writing, she is still waiting for approval while officials attempt to determine if "The World's Tallest File Cabinet" is too tall for its new location.[3]

Alvarez's sculpture succeeds as satire because in the early twenty-first century the file cabinet is associated with inefficiency. A file cabinet is no longer an exemplar of efficiency, rationalization, and speed. Instead, it represents the failure to save time and labor. A file cabinet's capacity embodies the facility of bureaucracies to produce paper, to delay, to always leave people in a state of want; as Ben Kafka argues, bureaucracy performs an important role in the modern world as an explanation for why people cannot always get what they want.[4] The integrity of a file cabinet, its metal construction, represents the impersonality of bureaucratic paperwork. Explaining bureaucratic impersonality, Kafka argues that paperwork has the character of inevitability. Procedure, the rational and impersonal way in which the arrangement of information on forms is used, always produces something new that replaces the authority of personal experience.[5] Therefore, a cabinet full of files also symbolizes the particular anxiety produced by the knowledge that paper records create an alternative paper-based reality to which officials defer.[6]

The file cabinet is an icon, symbolizing the opposite of what it once was, because it is still active. If you work for the federal government and you want to get paid in retirement, somebody 240 feet underground in Pennsylvania has to process your paperwork by hand, finding it in one of 28,000 file cabinets housed in a former limestone mine. These papers are so precious (and combustible) that hot lunches (pizza and four different deep-fried items) are delivered daily to the site's six hundred workers because open flames and toaster ovens are banned. Aboveground, the excess that constitutes bureaucracy-as-paperwork can be more visible. In 2012, the weight of file cabinets and paperwork threatened the structural integrity of a Veterans Benefits Administration office building in North Carolina. In this case, not even the file cabinet could contain paperwork, with a purported 37,000 files stacked two-feet high on top of the office's full file cabinets.[7]

Piles of Paper

Outside of government bureaucracies the failure of file cabinets to contain the paperwork an office produced came to represent a different set of problems for a different set of workers. In the 1970s, piles of paper became the main way to signify the new office phenomenon of "information overload." This was a popular uptake of a concept that originated in a midcentury merger of psychology and information systems theory. Scholars who initially wrote about information overload focused on the internal dynamics of information flow in social systems. They wanted to identify design flaws and redesign systems to make communication more efficient.[8]

According to Nick Levine, into the late 1980s, paper piled high was part of a "journalistic discourse about information overload [that] reflected the growth in the visibility of the stressed-out white collar worker overwhelmed by paper and his or her contradictory expectations to be at once a creative decision-maker and an information processor."[9] In this iteration of the alienated white-collar worker, the person doing some of the low-skill routine work that C. Wright Mills highlighted in his classic 1951 book has moved up within the office hierarchy. Managers and executives were now expected do some tasks that looked like clerical work; Mills's "enormous file" had become an "enormous pile."[10]

Piles of paper in offices hid the anxiety of executives forced to confront the "information processing" that had long been coded as women's work. Therefore, in the office, information overload was a product of a glitch in the gendered organization of office work. The arrival of the desktop computer accelerated this change. Promoted as a personal assistant, the computer left many high-level white-collar workers without someone to do their clerical work. However, the existence of piles of paper acknowledged that certain white-collar workers maintained sufficient status to ensure that, if they preferred not to, they could avoid this feminized labor without losing their jobs.

Piles of paper, kept to a manageable height and left on a desktop, could also be understood as exemplary information management practice. In the 1980s and 1990s, researchers who studied the organization of desks and offices argued that in the right hands a pile was a more efficient way to store and process information than a file cabinet. When papers were in a pile, a worker would find a document by looking down the edge of the pile, taking note of color, texture, thickness, and so on. The researchers described a number of uses for paper piles on desktops. Sometimes workers grouped papers by projects, with the most urgent projects closest to them and the most important documents placed at the top. On other occasions workers used piles to store ideas that could not be easily categorized or ideas that seemed important but had no immediate application. While such piles emerged from active, ongoing thinking, a dislike of formal classification also motivated workers; piles kept papers in a temporary holding pattern prior to filing. In summarizing their findings, a team of pile researchers who interviewed thirteen workers at Apple concluded, "Piling requires less mental effort."[11]

A Vertical Desktop

For pile researchers, studying the information management practices of Apple employees was an oddly appropriate choice. Research into how workers used the space of their desktops to organize paper and information was a response to the appearance of paper documents, manila folders, and file cabinets as icons on computer screens; those icons were part of an attempt to improve or replace the "desktop metaphor." In the 1970s, computer scientists used the idea of a desktop to help them develop the personal computer (the office had given metaphorical support to computers since the middle of the twentieth century, when "files" arrived in mainframe computers via punch cards, a collection of which had been named a file).[12] In 1984, Apple gave the desktop metaphor greater visibility with the launch of the Macintosh, which organized the user interface by representing a file as a paper document stored in a tabbed folder or discarded in a wastepaper basket. Apple's success meant that systems with similar interfaces followed on a "massive scale," cementing the desktop metaphor as the way to present the personal computer as a work space.[13]

Although early Macintosh desktops lacked a file cabinet icon, other visualizations of the desktop metaphor used such icons. In 1983, inspired by Apple's initial attempts to visualize the desktop metaphor on its expensive Lisa model, Commodore International introduced a ROM cartridge called Magic Desk I for its Commodore 64, a considerably less powerful but cheaper home computer. Instead of presenting a vertical desktop, it extended the metaphor to the space of an office to offer a more detailed desktop than anything Apple would ever do. The user was given a desk with drawers and a typewriter, calculator, telephone, account book, and Rolodex on top. A three-drawer file cabinet with a clock on top was placed next to the desk. Only the clock, typewriter, and file cabinet would do anything if selected by a user. To save a document, a user opened a file drawer by using a joystick to position a disembodied white hand with an extended index finger over the file drawer icon. Inside the drawer were ten yellow lines placed vertically in a list; each line had a tab to simulate a "folder" that a user could select and name.[14] Windows operating systems used file cabinet icons into the 1990s. Windows 3.1, released in 1992, included an icon of an open file drawer divided by tabbed folders; two documents partially removed from rear folders made sure that users understood what they were doing when they chose a location to save a document. Therefore, as an icon within a metaphor, the late twentieth-century file cabinet was as efficient as the early twentieth-century filing cabinet.

The triumph of the desktop metaphor was attributed to the way it provided "a set of intuitively clear underlying principles that rendered a consistent mental model of the digital workspace as a whole."[15] Its success also made clear the assumption that early personal computers would be office technologies like their larger-scale predecessors. However, the belief that such visualizations would prompt intuitive responses spoke to the naturalization of the modern office, not the neutrality of the metaphor. The gendered technologies and work practices of the office were no longer automatic like a machine but had become instinctive for people. Naturalized, these technologies apparently ensured that an office would process information productively.

Not surprisingly, this critique of the power dynamics that constituted the modern office did not drive the critique of the desktop metaphor in pile research. This research was about how to design a graphic user interface to more "accurately" represent how people managed information so workers could become more efficient and productive. The authors of one article on how people piled paper argued, "The folder as the sole container type presents an impoverished set of possibilities."[16] That is, the critique of the desktop metaphor came from the belief that information can rarely be placed into a single category.[17] This critique has a history as long as that of the desktop metaphor. In one of its more infamous formulations, Ted Nelson, the information technology pioneer who coined the terms *hypertext* and *hypermedia,* has argued that the use of paper simulations in computing is "like tearing the wings off a 747 and driving it as a bus on a highway."[18] For Nelson, the compartmentalization of files suggests that there is no overlap between the things people do.[19] In contrast, computers should be able to capture the reality of the overlaps; the affordances of paper should not restrict the possibilities of computers.

Although piles did not appear on computer screens, file cabinets did disappear from them. They disappeared when file icons and the logic of the desktop metaphor lost their monopoly on how people interacted with computers and information. In the early twenty-first century, cell phones and the "'Googlization' of the desktop" (which enabled keyword search) provided alternatives to the location-based search logic dependent on tabbed manila folders to represent storage.[20]

Once again the file cabinet appears as a distinctly twentieth-century technology. This dating underscores the filing cabinet as a response to a particular moment of capitalism when an increase in the volume of information occurred at a time when information was recorded, stored, and circulated primarily on paper. When capitalism changed its shape, the volume of information increased, and an alternative medium developed to record, store, and circulate information, the file cabinet struggled to perform the fundamental infrastructural work required to make information accessible.

Using Google to File

The effects of the changing shape of capitalism on information are illustrated, somewhat oddly, by a comic strip from 1921 that I described earlier. In it, a male file clerk decides to quit after his boss refuses to pay him more. Before he leaves, in an act of defiance, he removes papers from a filing cabinet and throws them around the

Figure A.2. Reprinted in an issue of *Filing and Office Management,* this 1921 cartoon about a male file clerk's ultimately successful attempt to get a pay raise features a clerk with the historically uncanny name of Mr. Google. Google is a family name, although the 1920 U.S. census recorded fewer than ten families with the name. Larry Page and Sergey Brin named their search engine after a common misspelling of *googol,* a number equal to 10 to the power of 100. Public domain/Google Books.

office. His boss then changes his mind, and the clerk gets the news of his pay increase while surrounded by loose papers and files.[21] I discovered this image five years into my research. It stood out because when the boss makes the decision to increase the file clerk's pay, he identifies the man as "Mr. Google." The coincidence of the name Google being associated with access to information was uncanny to say the least. The fact that the image I found online was marked as "digitized by Google" made it even weirder. While this coincidence borders on the ridiculous, it also illustrates that categories of information overload and information management take vastly different forms at different historical moments. At one time the filing cabinet was the symbol of orderly information management. By end of the twentieth century, Google search had taken that mantle. In the comic strip, Mr. Google sabotages the filing system by spilling papers out of their "proper" locations. Ironically, this results in exactly the kind of giant pile of papers (i.e., the web made up of pages) that Google the corporation provides access to through page rank. The chaos Mr. Google creates is exactly the chaos Google promises to manage. However, although different, both the filing cabinet and the search engine became organizing principles for the capitalist management of information.

Therefore, to argue for the historical specificity of the file cabinet is not to diminish its significance. Yes, the demise of the file cabinet is the demise of the monopoly of a particular way of organizing information that depended on the creation of a system into which information could be placed, the certainty that information had a proper place, and the confidence that such a place could be known in advance. But the move from piling unbound paper to storing it on its long edge in tabbed manila folders in drawers in a cabinet radically changed how people accessed papers. Further, it gave information a metaphorical structure, an organizing principle. The filing cabinet shaped a way of understanding and knowing. Designed by the ideals of efficiency and the power dynamics of gender and labor, it turned information into a thing that could be seen to exist at the ends of people's fingertips.

Acknowledgments

This book took longer to write than I planned. After a while, I started saying that writing the book had become an act of resistance to the ideas of efficiency instantiated in the filing cabinet. But that was not quite true. The reality is that I think and write slowly. The slowness in thinking and writing came from taking on a project that was initially smarter than I was. The only frustration came when I had to put it on the back burner for a few years after I was drafted into administrative work to build a new program.

There was something positive in taking a long time to complete this book: it gave me time to talk to a lot of very smart people about the history of the filing cabinet. Three people deserve particular thanks for countless conversations and readings of drafts over the years. Dylan Mulvin talked filing cabinets in Cambridge and London, read the complete manuscript in a very rough draft, and suggested the book's subtitle. Deidre Lynch's perceptive insights and good humor pushed me to engage more fully with the larger stakes involved in this project. As he has for more than a dozen years, Josh Lauer traveled down from New Hampshire to allow me to talk incessantly about my work, all the while offering gentle critiques.

As I revised the book, Mar Hicks, Jeremy Packer, and Liam Young helped me better understand debates and arguments I couldn't find my way through on my own. Although I did not call on them to rescue me at the end of this project, my ongoing conversations with friends Jack Braitch, Lisa Gitelman, James Hay, and Jonathan Sterne greatly improved this book. It is also immeasurably better because

Rachel Dubrofsky, Alan Liu, Lynn Spigel, and Martin Parker read and commented on drafts of articles and chapters.

Over the years I had great conversations with many people, sometimes during or immediately after conference sessions, but usually in the more relaxed environments of bars and restaurants. In some cases we might have had only one conversation about the book, but trust me, it was important! Thank you to Mark Andrejevic, Tony Ballantyne, Sarah Banet-Weiser, Matthew Battles, Ann Blair, Grant Bollmer, Nick Couldry, Alex Cszair, Anne Cullen, Michael Denning, Marco Deseriis, Paul Duguid, Lori Emerson, Anna Feigenbaum, Megan Finn, Melissa Fitzgerald, Radhika Gajjala, Kelly Gates, Ron Greene, Mel Gregg, Mark Hayward, Dave Hesmondhalgh, Hansun Hsiung, Matt Hull, Louis Hyman, Diana Kamin, Matt Kirschenbaum, Kristian Kloeckl, Julilly Kohler-Hausman, Markus Krajewski, Marie Leger, Jen Light, Shannon Mattern, Cait McKinney, Anna-Maria Meister, Alex Monea, Dave Parisi, Victor Pickard, Mary Poovey, Leah Price, Josh Reeves, Carrie Rentschler, Ale Renzi, Andrew Ross, Andy Russell, Sarah Sharma, Josh Shepherd, Tom Streeter, Ted Striphas, Lana Swartz, Tamara Plakins Thornton, Nanna Bonde Thylstrup, Christine von Oertzen, Haidee Wasson, Moira Weigel, Jason Wilson, Stacy Wood, Shaun Wortis, JoAnne Yates, Mike Zundel, and the late Edith Ackerman. In addition to talking about this project, Kevin Bruyneel, Kirsten Weld, Marina Leslie, Murray Forman, and Joanne Morreale provided great friendship in my hometown.

My thinking about this book was particularly helped by discussions at Microsoft New England and the Max Planck Institute for the History of Science (thank you to Nancy Byam and Hansun Hsiung for the invitations) and with the audience at the Spatial Organization of Media panel at the Society for Cinema and Media Studies conference in 2015 (and my fellow panelists, Matt Kirschenbaum and Shannon Mattern). Thank you to Tony Ballantyne and the history department at the University of Otago for hosting me during the first half of 2013 when I started to work on this project.

As I was struggling with what kind of history of the filing cabinet to write, the insights and support of Susan Ferber and Sean Cubitt were particularly encouraging. When I did figure that out, I was very happy that the University of Minnesota Press agreed to publish this book; the mix of history and theory I have written fits with the strengths associated with its publishing program. Pieter Martin has been a fantastic and patient editor—the book is infinitely better for his editorial skill. I also appreciate the reviewers he found and the critiques and suggestions that Ben Kafka and Shannon Mattern offered on a draft that was much rougher than I recognized when I submitted it. At the University of Minnesota Press I am very grateful to Laura Westlund, Rachel Moeller, and Anne Carter for all their work getting this book from manuscript to production in the very unusual and uncertain circumstances of 2020. In copyediting the manuscript Judy Selhorst did a great job bringing my prose to life. Thanks to Doug Easton for his work on the index.

In a book that is about the infrastructure critical to paperwork, to how paper works, I want to thank the people and institutions that provided specific infrastructural support for the writing of this book: Northeastern University's College of Arts, Media, and Design, which gave financial support for me to complete research and attend conferences; the staff at Eye Q in Harvard Square, who convinced me to get another set of progressive lenses specifically designed so I could look at a screen for hours on end without getting headaches; Nick Marini, who used his impressive Bose staff discount to allow me to afford noise-canceling headphones; Lisa Pierce, who helped me rethink how I approach writing; and Harvard University, which allows spouses of employees to use its libraries (with only a handful of annoying limitations). Finally, but most important, thank you to Dale Herbeck, who for the past half dozen years has been the ideal department chair, providing a buffer between faculty and excessive administrative demands as well as ensuring that the research and teaching of his faculty get the respect they deserve.

This book would not exist without the work of archivists and local historical societies who helped me find the records and publications of filing equipment companies. The irony was not lost on me that in many cases the records of companies devoted to the storage of records could not be located. Therefore, I am particularly grateful to the Fenton Historical Society and the Herkimer County Historical Society for saving the records of Art Metal and the Library Bureau, respectively, and to Steelcase for maintaining a corporate archive. I am very happy to acknowledge the expertise and work of Linda Gross and Angela Schad (Hagley Museum and Library), Sue Perkins (Herkimer County Historical Society), Karen Livsey and Victoria Parker (Fenton Historical Society), Kathy Reagan and Justine Bailey

(Steelcase), Trina Brown (Smithsonian Libraries), Beryl Gabel (Lakeshore Museum Center), Heather Tucker (TOPS Products), and the staff at Grand Rapids Public Library, Grand Rapids Public Museum, the Baker Library, and the Interlibrary Loan Department at Northeastern University. My research benefited from the generosity of JoAnne Yates and Tom Hench, who sent me photocopies, catalogs, and ephemera they had collected during their research on the history of office equipment.

Finally, I offer special thanks to family: the Shackelfords (Jim, Janet, Sarah, Jack, Ryan, Rhiana, Emmett, and Wyatt) and, spread across the world, Stephen, Delwyn, and Cleo; Grant and Alf; Lisa and Campbell, and my mother, Yvonne, a retired self-taught archivist who seemed to enjoy the minutiae of the filing cabinet almost as much as I did. This book is dedicated to Erin and Edie, who in their everyday lives provide ample evidence of what matters most in this world. Edie was four years old when I started research for this book; now she is a thirteen-year-old climate activist who finds time to remind me that I should mention that in *Rogue One* the data facility that stores the plans for the Death Star is a great example of verticality and information. While it probably seems to Erin that I have been working on this book for two-thirds of her life, I am truly grateful for the time, space, encouragement, and humor she gave me to complete it. That being said, I am sure Erin and Edie both agree it is definitely time to file it.

Notes

Preface

1. Craig Robertson, *The Passport in America: The History of a Document* (New York: Oxford University Press, 2010).
2. Quoted in Morton Keller, *Affairs of State: Public Life in Late Nineteenth Century America* (Cambridge, Mass.: Harvard University Press, 1977), 318.
3. Milton Gustafson, "State Department Records in the National Archives: A Profile," *Prologue* 2 (1970): 179.
4. JoAnne Yates, *Control through Communication: The Rise of System in American Management* (Baltimore: Johns Hopkins University Press, 1989).
5. Geertz notes that the double entendre of "cock" exists in Balinese just as it does in English. Clifford Geertz, "Deep Play: Notes on the Balinese Cockfight," in *The Interpretation of Cultures: Selected Essays* (New York: Basic Books, 1973), 417. Although Geertz seeks to put details into social context, I am using a common misreading that defines "thick description" in a more narrow sense.
6. Cornelia Vismann, *Files: Law and Media Technology,* trans. Geoffrey Winthrop-Young (Stanford, Calif.: Stanford University Press, 2008).

Introduction

1. Terrence L. Uber, "Creating the Steel Chapel: A Study of Commercial Office Furniture Design in the United States from 1876–1925" (PhD diss., Case Western Reserve University, 2001), 258.
2. John Tagg, *The Disciplinary Frame: Photographic Truths and the Capture of Meaning* (Minneapolis: University of Minnesota Press, 2008), 209–33; Allan Sekula, "The Body and the Archive," *October* 39 (Winter 1986): 3–65; Josh Lauer, *Creditworthy: A History of Consumer Surveillance and Financial Identity in America* (New York: Columbia University Press, 2017). For analyses that foreground the materiality of filing systems, see JoAnne Yates, "From Press Book and Pigeonhole to Vertical Filing: Revolution in Storage and Access Systems for Correspondence," *Journal of Business Commu-*

nication 19, no. 3 (1982): 5–26; Vismann, *Files*; Bruno Latour, *The Making of Law: An Ethnography of the Conseil d'Etat,* trans. Marina Brilman and Alain Pottage (Cambridge: Polity Press, 2010); Markus Krajewski, *Paper Machines: About Cards and Catalogs, 1548–1929,* trans. Peter Krapp (Cambridge, Mass.: MIT Press, 2011); Shannon Mattern, "Indexing the World of Tomorrow: How the 1939 World's Fair Anticipated Our Current Obsession with Urban Data Science and 'Smart' Cities," *Places Journal* (February 2016), https://placesjournal.org.

3 Gerri Flanzraich, "The Role of the Library Bureau and Gaylord Brothers in the Development of Library Technology, 1876–1930" (PhD diss., Columbia University, 1990), 350–64; Uber, "Creating the Steel Chapel," 130–62, 240–88; Alexandra Lange, "White Collar Corbusier: From the *Casier* to the *Cités d'affaires,*" *Grey Room,* no. 9 (Fall 2002): 58–79. My research benefited from this groundbreaking scholarship.

4 Geoffrey Bowker and Susan Leigh Starr, *Sorting Things Out: Classification and Its Consequences* (Cambridge, Mass.: MIT Press, 1999); Lisa Parks and Nicole Starosielski, eds., *Signal Traffic: Critical Studies of Media Infrastructure* (Urbana: University of Illinois Press, 2015).

5 Max Weber, *Economy and Society* (Berkeley: University of California Press, 1978), 957. See also Daniel Nelson, *Managers and Workers: Origins of the Twentieth-Century Factory System in the United States, 1880–1920,* 2nd ed. (Madison: University of Wisconsin Press, 1995).

6 On affordances, see Abigail J. Sellen and Richard H. R. Harper, *The Myth of the Paperless Office* (Cambridge, Mass.: MIT Press, 2002), 14–16. As I discuss in the following section, affordance is one of several concepts I use to think about objects in everyday life.

7 Although scholars (including information historians) frequently use information and knowledge interchangeably it is important to separate the two. For examples of the latter, see John Seely Brown and Paul Duguid, *The Social Life of Information* (Cambridge, Mass.: Harvard Business School Press, 2002), 119–20; Geoffrey Nunberg, "Farewell to the Information Age," in *The Future of the Book,* ed. Geoffrey Nunberg (Berkeley: University of California Press, 1996), 103–38.

8 Nick Couldry, "Recovering Critique in an Age of Datafication," *New Media & Society* 22, no. 7 (July 2020): 1140.

9 Ronald E. Day, *The Modern Invention of Information: Discourse, History, and Power* (Carbondale: Southern Illinois University Press, 2001), 12. See also Rob Kitchin, "Big Data, New Epistemologies and Paradigm Shifts," *Big Data & Society* 1, no. 1 (April–June 2014), https://doi.org/10.1177/2053951714528481.

10 American Institute of Filing, *A Course in Correspondence Filing for Home Study* (Boston: Library Bureau, 1921), 13, Library Bureau Papers, Herkimer County Historical Society, Herkimer, New York (hereafter Herkimer County Historical Society).

11 Irene Warren, Marion C. Lyons, and Frank C. McClelland, *Filing and Indexing with Business Procedure* (Chicago: Rand McNally, 1924), 46.

12 Alex F. Osborn and Robert E. Ramsay, *The Optimism Book for Offices* (Jamestown, N.Y.: Art Metal Construction Co., 1919), 40.
13 Anke te Heesen, *The World in a Box: The Story of an Eighteenth-Century Picture Encyclopedia,* trans. Ann M. Hentschel (Chicago: University of Chicago Press, 2002); Glenn Adamson, "The Labor of Division: Cabinetmaking and the Production of Knowledge," in *Ways of Making and Knowing: The Material Culture of Empirical Knowledge,* ed. Pamela Smith, Amy Myers, and Harold Cook (Ann Arbor: University of Michigan Press, 2014), 243–79.
14 For a relevant history of the Library Bureau, see Gerri Flanzraich, "The Library Bureau and Office Technology," *Libraries and Culture* 28, no. 4 (Fall 1993): 403–29.
15 "History of Vertical Filing," supplement, *L.B. Monthly News,* no. 24 (December 1916): 48; Estelle Hunter, *Modern Filing Manual* (Rochester, N.Y.: Yawman and Erbe, 1941), 6. In the 1930s, Allen Chaffee claimed that the filing cabinet exhibited at the 1893 Chicago World's Fair had received a gold medal. Allen Chaffee, *How to File Business Papers and Records: A Practical Business Manual Dealing with Filing Systems and Equipment in Use Today* (New York: McGraw-Hill, 1938), 4. No previous retelling of the Library Bureau origin story had suggested this. Nor has any subsequent archival evidence been found to support his claim. However, the catalog for the Chicago fair suggests that a filing cabinet was exhibited there. The catalog states that the Library Bureau's exhibits included a "card-case for records of charitable societies." Quoted in "Antique Filing Cabinets," Early Office Museum, accessed September 3, 2020, http://www.officemuseum.com/filing_equipment_cabinets.htm. Given the origin of this "card case" in a charitable society's need to store records and the language used to describe cabinets at the time, it is likely this is a reference to what would by the end of the decade be called a vertical filing cabinet.
16 "Edwin G. Seibels: Legacy of Leadership Profile," Knowitall, 1999, https://www.knowitall.org; Seibels Bruce Insurance Companies, *Seibels Bruce since 1869* . . . (Columbia, S.C.: Seibels Bruce Insurance Companies, n.d.), 9.
17 For a list of patents issued to office equipment companies for vertical filing cabinets in the period 1900–1924, see Flanzraich, "Role of the Library Bureau," 354a–d. On the difficulties involved in pursuing a patent infringement claim, see Library Bureau, "Boston Salesmen's Periodical Meetings 1898–1899," 227, Herkimer County Historical Society. For examples of patent disputes that made it to court, see Globe-Wernicke v. Fred Macey Co., 119 F. 695 (1902); Macey Co. v. Globe-Wernicke Co., 180 F. 401 (1910).
18 "The Key on the Cover," *Office Appliances,* November 1921, 13.
19 "Remington Rand Plans Outlined," *Wall Street Journal,* February 12, 1927, 7. On the development of the office equipment industry, see Uber, "Creating the Steel Chapel."
20 Shaw-Walker, *Built Like a Skyscraper* (Muskegon, Mich.: Shaw-Walker,

1927), 4–9, Trade Catalog Collection, Imprints, Hagley Museum and Library, Wilmington, Delaware (hereafter Hagley Museum and Library).

21 Liam Cole Young, *List Cultures: Knowledge and Poetics from Mesopotamia to BuzzFeed* (Amsterdam: Amsterdam University Press, 2017), 41.

22 Shannon Mattern, "Intellectual Furnishing: The Physical and Conceptual Architectures of Our Knowledge Institutions," Words in Space, 2014, https://wordsinspace.net/shannon/2014/02/26/7542.

23 I include aspects of critical infrastructure, media archaeology, cultural techniques, science studies, affordances, and McLuhan-esque media analysis in "media materialism." However, as discussed in the Preface, I borrow ideas and concepts while remaining skeptical of the bracketing of power dynamics in many of these theoretical frameworks: Jussi Parikka, *What Is Media Archaeology?* (Cambridge: Polity Press, 2012); Bernhard Siegert, *Cultural Techniques: Grids, Filters, Doors, and Other Articulations of the Real*, trans. Geoffrey Winthrop-Young (New York: Fordham University Press, 2015); Geoffrey Winthrop-Young, "Material World: Geoffrey Winthrop-Young Talks with Bernhard Siegert," *Artforum International*, Summer 2015, 324–33; Bruno Latour, "Visualization and Cognition: Thinking with Eyes and Hands," *Knowledge and Studies in the Sociology of Culture Past and Present* 6 (1986): 1–40; Sarah Sharma, "It Changes Space and Time! Introducing Power-Chronography," in *Communication Matters: Materialist Approaches to Media, Mobility and Networks*, ed. Jeremy Packer and Stephen B. Crofts Wiley (London: Routledge, 2012), 66–77; Parks and Starosielski, *Signal Traffic*.

24 As John Durham Peters notes in a survey of debates about technological determinism: "Whenever the patently obvious must be invoked solemnly and predictably, thinking has stopped." John Durham Peters, "Why We Use Pencils and Other Thoughts on the Archive," in *Media History and the Archive*, ed. Craig Robertson (London: Routledge, 2011), 113.

25 Grant Bollmer, *Materialist Media Theory: An Introduction* (New York: Bloomsbury, 2019), 12. This understanding of media materialism draws from feminist technology studies. See Sarah Sharma and Rianka Singh, eds., *MsUnderstanding Media* (Durham, N.C.: Duke University Press, forthcoming).

26 Other scholars have highlighted this moment to separate the history of modern information from the history of computing; see, for example, Jorge Reina Schement, "Porat, Bell, and the Information Society Reconsidered: The Growth of Information Work in the Early Twentieth Century," *Information Processing and Management* 26, no. 4 (1990): 449–65; Alistair Black, Dave Muddiman, and Helen Plant, eds., *The Early Information Society: Information Management in Britain before the Computer* (Aldershot, England: Ashgate, 2007).

27 Ann Blair, *Too Much to Know: Managing Scholarly Information before the Modern Age* (New Haven, Conn.: Yale University Press, 2011). Blair notes the "impersonal and controlled processing of information" that made

early twentieth-century modes of information organization different from those of the early modern period (265).

28 Forrest E. Cardullo, "Industrial Administration and Scientific Management—1," *Machinery,* July 1912, 846. Cardullo's article is an example of the ease with which writers applied different names to the instrumental conception of information. While he introduced it as "accurate information," Cardullo also labeled the content "data" and the result "knowledge," albeit making clear this was a very particular form of knowledge.

29 The filing cabinet contained what Warren Weaver refers to as information in its "ordinary usage"—that is, information that maintains its association with meaning, as distinct from the information of information theory. Claude Shannon and Warren Weaver, *The Mathematical Theory of Communication* (Urbana: University of Illinois Press, 1948), 8. For arguments about the changing meanings of information from the late eighteenth century onward, see Nunberg, "Farewell to the Information Age"; Paul Duguid, "The Ageing of Information: From Particular to Particulate," *Journal of the History of Ideas* 76, no. 3 (July 2015): 347–68.

30 Angel Kwolek-Folland, *Engendering Business: Men and Women in the Corporate Office 1870–1930* (Baltimore: Johns Hopkins University Press, 1994), 109.

31 Yehouda Shenhav, *Manufacturing Rationality: The Engineering Foundations of the Managerial Revolution* (New York: Oxford University Press, 1999); David F. Noble, *America by Design: Science, Technology, and the Rise of Corporate Capitalism* (New York: Alfred A. Knopf, 1977), 259–63.

32 James Beniger argues for the importance of information to control and planning. In summarizing Beniger's work, John Durham Peters highlights the idea that "information is an entity that becomes salient in a certain kind of society—a society of great speed, complexity, and size." While this overlaps with my larger argument about the relationship between information and efficiency, Beniger's more ambitious project to make information processing the basis of human society removes information from the power dynamics critical to my arguments in this book. As Peters puts it, Beniger's *The Control Revolution* is a "brilliant execution of a flawed project." John Durham Peters, "The Control of Information," *Critical Review* 1, no. 4 (Fall 1987): 5–6; James Beniger, *The Control Revolution: Technological and Economic Origins of the Information Society* (Cambridge, Mass.: Harvard University Press, 1986). See also Alfred Chandler, *The Visible Hand: The Managerial Revolution in American Business* (Cambridge, Mass.: Belknap Press, 1976).

33 Andrew Piper, *Book Was There: Reading in Electronic Times* (Chicago: University of Chicago Press, 2012), 60.

34 Lisa Gitelman, *Paper Knowledge: Toward a Media History of Documents* (Durham, N.C.: Duke University Press, 2014), 4. Gitelman is summarizing Geoffrey Nunberg's argument in "Farewell to the Information Age." For other examples, see Krajewski, *Paper Machines*; Anke te Heesen, *The*

Newspaper Clipping: A Modern Paper Object, trans. Lori Lantz (Manchester: Manchester University Press, 2014).

35 Alex Csiszar, "Bibliography as Anthropometry: Dreaming Scientific Order at the Fin de Siècle," *Library Trends* 62, no. 2 (2013): 442–43.

36 Staffan Müller-Wille, "Names and Numbers: 'Data' in Classical Natural History, 1758–1859," *Osiris* 32, no. 1 (2017): 115. See also Christine von Oertzen, "Machineries of Data Power: Manual versus Mechanical Census Compilation in Nineteenth-Century Europe," *Osiris* 32, no. 1 (2017): 129–50.

37 Quoted in W. Boyd Rayward, introduction to *International Organization and Dissemination of Knowledge: Selected Essays of Paul Otlet,* by Paul Otlet, trans. and ed. W. Boyd Rayward (Amsterdam: Elsevier Science, 1990), 1. See also Bernd Frohmann, "The Role of Facts in Paul Otlet's Modernist Project of Documentation," in *European Modernism and the Information Society: Informing the Present, Understanding the Past,* ed. W. Boyd Rayward (Aldershot, England: Ashgate, 2008), 75–88; W. Boyd Rayward, "Visions of Xanadu: Paul Otlet (1868–1944) and Hypertext," *Journal of the American Society for Information Science* 45, no. 4 (1994): 235–50; Day, *Modern Invention of Information,* 9–21.

38 L. S. Jast, "The Commercial Library," *Library Association Record* 19 (1917): 122. See also Alistair Black, "From Reference Desk to *Desk Set*: The History of the Corporate Library in the United States and the UK before the Adoption of the Computer," in *Best Practices for Corporate Libraries,* ed. Sigrid E. Kelsey and Marjorie J. Porter (Santa Barbara, Calif.: Libraries Unlimited, 2011), 3–24; Robert Williams, "Documentation and Special Libraries Movements in the United States, 1910–1960," *Journal of the American Society for Information Science* 48, no. 9 (1997): 775–81.

39 Jeff Johnson, "How Faithfully Should the Electronic Office Simulate the Real One?," *ACM SIGCHI Bulletin* 19, no. 2 (1987): 21.

40 Vismann, *Files,* 161–64.

41 Wolfgang Ernst, *Digital Memory and the Archive* (Minneapolis: University of Minnesota Press, 2013).

42 Ernst, 120.

43 I am talking about the file folder icons that appear on computer screens. As Vismann shows, the file as an abstraction used by programmers and engineers can be connected to earlier moments in the history of files and registers. Vismann, *Files,* 71–101.

44 Lewis Mumford, *Technics and Human Development* (New York: Harcourt Brace Jovanovich, 1966), 141.

45 For an insightful analysis of containers that uses Mumford's argument, see Zoe Sofia, "Container Technologies," *Hypatia* 15, no. 2 (Spring 2000): 181–201. For an equally insightful discussion about containers and containment, see Sina Najafi and Gerhard Wolf, "Thinking inside the Box: An Interview with Gerhard Wolf," *Cabinet* 60 (Winter/Spring 2015–16): 69–77.

46 My argument about retrieval borrows some of the language (but not the argument) of Hartmut Winkler's "Geometry of Time: Media, Spatialization, and Reversibility" (paper presented at the conference "Media Theory on the Move," Potsdam, May 21–24, 2009), http://homepages.uni-paderborn.de/winkler.

47 C. Wright Mills, *White Collar: The American Middle Classes* (New York: Oxford University Press, 1951), 189.

48 Harry Braverman, *Labor and Monopoly Capitalism: The Degradation of Work in the Twentieth Century*, 25th anniversary ed. (New York: Monthly Review Press, 1998).

Chapter 1. Verticality

1 For a provocative take on verticality and its relationship to capitalism, see the following series of articles by Martin Parker: "Organizing: Skyscrapers and Multitudes," *Critical Perspectives on International Business* 3, no. 3 (2007): 220–22; "Skyscrapers: The City and the Megacity," *Theory, Culture & Society* 31, nos. 7–8 (2014): 267–71; "Vertical Capitalism: Skyscrapers and Organization," *Culture and Organization* 21, no. 3 (2015): 217–34; "Tower Cranes and Organization Studies," *Organization Studies* 38, no. 7 (2017): 989–1004. For a more general critique of the vertical and the social, see Stephen Graham, *Vertical: The City from Satellites to Bunkers* (London: Verso, 2016).

2 William Leffingwell, *Textbook of Office Management* (New York: McGraw-Hill, 1932), 201.

3 William Henry Leffingwell, *Office Management: Principles and Practices* (New York: A. W. Shaw, 1925), 416.

4 In contrast, Alexandra Lange argues that the filing cabinet was a "building block for the twentieth-century design of offices." Lange, "White Collar Corbusier," 59. She makes that argument as part of a project that seeks to historicize Le Corbusier by placing his work within a debate about the politics of organization. As I note later in this chapter, the importance placed on movement in the organization of an office is critical to how verticality is deployed in the office.

5 For a longer history of the vertical and organization, see Martin Parker, "Angelic Organization: Hierarchy and the Tyranny of Heaven," *Organization Studies* 30, no. 11 (2009): 1281–99.

6 Reinhold Martin notes that the vertical subordinates parts to the whole. See Reinhold Martin, *The Organizational Complex: Architecture, Media, and Corporate Space* (Cambridge, Mass.: MIT Press, 2003), 23.

7 Alexandra Lange, "Tower, Typewriter and Trademark: Architects, Designers and the Corporate Utopia, 1956–1964" (PhD diss., New York University, 2005).

8 Giovanni Arrighi, *The Long Twentieth Century: Money, Power and the Origins of Our Times* (London: Verso, 2010), 290–91.

9 Parker, "Organizing," 221.

10 Mills, *White Collar*, 209. Although Mills alludes to cross-referencing ("interlinked"), as I argue in chapter 5 cross-referencing was an awkward add-on to the vertical filing cabinet, which was designed to store closed systems such as alphabetically ordered files.
11 J. Carson Webster, "The Skyscraper: Logical and Historical Considerations," *Journal of the Society of Architectural Historians* 18, no. 4 (December 1959): 136; Parker, "Vertical Capitalism," 233.
12 Webster, "The Skyscraper."
13 Webster, 128.
14 Gail Fenske, *The Skyscraper and the City: The Woolworth Building and the Making of Modern New York* (Chicago: University of Chicago Press, 2008), 167. For a detailed discussion of the construction process, see Fenske, 166–215.
15 Fenske, 4.
16 George H. Douglas, *Skyscrapers: A Social History of the Very Tall Building in America* (Jefferson, N.C.: McFarland, 2004), 11–15; Thomas J. Misa, *A Nation of Steel: The Making of Modern America 1865–1925* (Baltimore: Johns Hopkins University Press, 1995), 63–89.
17 Misa, *Nation of Steel*, 87.
18 Fenske, *Skyscraper and the City*, 25.
19 Parker, "Vertical Capitalism," 224.
20 Roland Marchand, *Advertising the American Dream* (Berkeley: University of California Press, 1985), 242.
21 David Nye, *American Technological Sublime* (Cambridge, Mass.: MIT Press, 1994), 96.
22 Parker, "Vertical Capitalism," 218.
23 Keith D. Revell, "Law Makes Order: The Search for Ensemble in the Skyscraper City, 1890–1930," in *The American Skyscraper: Cultural Histories*, ed. Roberta Moudry (New York: Cambridge University Press, 2005), 46.
24 Lutz Hartwig, "Lifts and Architecture: Lift Architecture," in *Vertical: Lift, Escalator, Paternoster—A Cultural History of Vertical Transport*, ed. Vittorio Magnago Lanpugnani and Lutz Hartwig (Berlin: Ernst & Sohn, 1994), 44.
25 Quoted in Parker, "Vertical Capitalism," 226.
26 Louis Sullivan, "The Tall Office Building Artistically Considered," *Lippincott's Monthly Magazine*, March 1896, 404.
27 Frank Lloyd Wright, *The Future of Architecture* (New York: Horizon Press, 1953), 166, emphasis added.
28 Wright, 164, emphasis added.
29 Wright, 168, 171, 162.
30 Wright, 22.
31 Revell, "Law Makes Order," 41.
32 Revell, 41.
33 Hartwig, "Lifts and Architecture," 44.
34 Fenske, *Skyscraper and the City*, 239–40.

35 The Woolworth Building's connections to subway lines represented an attempt to manage the flow of workers outside the building.
36 Andreas Bernard, *Lifted: A Cultural History of the Elevator*, trans. David Dollenmayer (New York: New York University Press, 2014), 55, 59.
37 Siegert, *Cultural Techniques*, 97–120; Anna-Maria Therese Meister, "From Form to Norm: Systems and Values in German Design circa 1922, 1936, 1953" (PhD diss., Princeton University, 2018), 7.
38 Hannah B Higgins, *The Grid Book* (Cambridge, Mass.: MIT Press, 2008), 6.
39 Siegert, *Cultural Techniques*, 120.
40 Higgins, *Grid Book*, 3.
41 Higgins, 4.
42 Albert Kahn, "Designing Modern Office Buildings," *Architectural Forum* 52 (June 1930): 775.
43 Carol Willis, *Form Follows Finance: Skyscrapers and Skylines in New York and Chicago* (New York: Princeton Architectural Press, 1995), 26.
44 Earle Shultz and Walter Simmons, *Offices in the Sky* (Indianapolis: Bobbs-Merrill, 1959), 129.
45 Nikil Saval, *Cubed: A Secret History of the Workplace* (New York: Doubleday, 2014), 110.
46 Willis, *Form Follows Finance*, 28–29.
47 R. H. Shreve, "The Empire State Building Organization," *Architectural Forum* 52 (June 1930): 772.
48 R. H. Shreve, "The Economic Design of Office Buildings," *Architectural Record*, April 1930, 352, 355.
49 As Alexandra Lange argues, the straight-line designs of a factory did not have an impact on the design of office buildings until the middle of the twentieth century, when some businesses began moving their headquarters out of cities. Lange, "Tower, Typewriter and Trademark," 36–39.
50 L. C. Walker, *The Office and Tomorrow's Business* (New York: Century, 1930), 96.
51 Lange offers a useful analysis of work flow in 1920s offices that acknowledges the impact of office furniture. However, she leaves too implicit the fact that very few offices followed through on straight-line designs, hence the filing cabinet did not play the significant role in the design of offices and the placement of furniture and machines in offices that Lange suggests. Lange, "White Collar Corbusier," 61–66.
52 Eugene Benge, *Cutting Clerical Costs* (New York: McGraw-Hill, 1931), 59.
53 Walker, *Office and Tomorrow's Business*, 47.
54 Roberta Moudry, "The Corporate and the Civic: Metropolitan Life's Home Office Building," in Moudry, *American Skyscraper*, 131–32.
55 For discussion of the limited uptake of scientific management in offices, see chapter 4 of this book as well as Thomas Haigh, "Technology, Information, and Power" (PhD diss., University of Pennsylvania, 2003), 58–94.
56 Thomas Hine, "Office Intrigues: The Interior Life of Corporate Culture," in *On the Job: Design and the American Office*, ed. Donald Albrecht and

Chrysanthe B. Broikos (New York: Princeton Architectural Press, 2000), 138; Alexander Klose, *The Container Principle: How a Box Changes the Way We Think,* trans. Charles Marcum II (Cambridge, Mass.: MIT Press, 2009), 255–56; Stanley Abercrombie, "Office Supplies: Evolving Furniture for the Evolving Workplace," in Albrecht and Broikos, *On the Job,* 82, 85.

57 Uber, "Creating the Steel Chapel," 58–88.
58 Lee Galloway, *Office Management: Its Principles and Practices* (New York: Ronald Press, 1919), 91.
59 Leffingwell, *Office Management,* 391–409.
60 Hunter, *Modern Filing Manual,* 20.
61 *Wooton Patent Desks: A Place for Everything and Everything in Its Place* (exhibition catalog) (Indianapolis: Indiana State Museum, 1983).
62 Brian P. Luskey, *On the Make: Clerks and the Quest for Capital in Nineteenth-Century America* (New York: New York University Press, 2010); Michael Zakim, *Accounting for Capitalism: The World the Clerk Made* (Chicago: University of Chicago Press, 2018).
63 Carl C. Parson, *Office Organization and Management* (Chicago: La Salle Extension University, 1921), 164.
64 Louise Krause, *The Business Library: What It Is and What It Does* (San Francisco: Journal of Electricity, 1921), 11. Krause focuses on the business library, which was another response to the problem of information need and accumulation in the modern office. For more on this topic, see chapter 6.
65 Quoted in Osborn and Ramsay, *Optimism Book for Offices,* 37.
66 Amy Weaver, *Office Organization and Practice* (New York: Ginn, 1923), 86.
67 Leffingwell, *Office Management,* 394.
68 W. H. Leffingwell, "Nine Improvements on 'The Way It Was Always Done,'" *System,* September 1922, 318.
69 Warren et al., *Filing and Indexing with Business Procedure,* 39.
70 Hunter, *Modern Filing Manual,* 17–18.
71 Shaw-Walker, *Built Like a Skyscraper,* 6.
72 Warren et al., *Filing and Indexing with Business Procedure,* 53.
73 Art Metal Construction Co., *Art Metal Filing Equipment* (Jamestown, N.Y.: Art Metal Construction Co., n.d.), B6, Art Metal Records, Fenton Historical Society, Jamestown, New York (hereafter Fenton Historical Society).
74 Automatic File & Index Co., *Automatic Five Drawer Files* (Chicago: Automatic File & Index Co., n.d.), n.p., Smithsonian Trade Catalog Collection, National Museum of American History, Washington, D.C. (hereafter National Museum of American History).
75 Shaw-Walker, *Built Like a Skyscraper,* 7, 49.
76 For a critique of high heels and the politics of elevation and platforms, see Rianka Singh and Sarah Banet-Weiser, "Sky High: Platforms and the Feminist Politics of Visibility," in Sharma and Singh, *MsUnderstanding Media.*
77 "Organization and Management of a Centralized File Bureau," *The File,*

December 1935, 7, Herkimer County Historical Society; Edith Fulton, "Lookers and Finders," *The File,* November 1937, 8, Herkimer County Historical Society.
78 W. H. Leffingwell, "The Versatile Vertical File," *Office Economist,* January 1923, 14.
79 "Efficiency Helped Win Girl Promotion. Used Her Eyes to Observe Lost Motion Waste," *Idaho Statesman,* June 27, 1917, 10.
80 Lauer, *Creditworthy,* 121.
81 Editorial, *Filing,* June 1921, 8.
82 Bertha M. Weeks, *How to File and Index* (New York: Ronald Press, 1937), 156.
83 D. E. Hunter, Expansible Filing System, U.S. Patent 722,709, filed September 4, 1902, and issued March 17, 1903.
84 Library Bureau, *Proven Floor Plans for Counterheight Units* (Boston: Library Bureau, n.d.), 10, Herkimer County Historical Society.
85 "Unusual Counter-Height Installation," *LB File,* March 1924, 46, Herkimer County Historical Society.
86 "Key on the Cover," 13.

Chapter 2. Integrity

1 Day, *Modern Invention of Information,* 16–17. On Otlet's monographic principle, see Frohmann, "Role of Facts in Paul Otlet's Modernist Project."
2 Day, *Modern Invention of Information,* 19.
3 For a European example, see Krajewski, *Paper Machines,* 132.
4 Piper, *Book Was There,* 11.
5 Walter J. Ong, *Ramus: Method, and the Decay of Dialogue* (Cambridge, Mass.: Harvard University Press, 1958).
6 Krajewski, *Paper Machines,* 137.
7 William A. Vawter, "Origin and Development of Loose Leaf," *Office Appliances,* May 1917, 19.
8 In the early twentieth century, index cards were also used as an alternative to bound books. I limit my focus to loose-leaf ledgers because these were the basis of court cases that provide the foundation for my discussion of integrity.
9 Ferdinand Risque, *Loose Leaf Books and Systems for General Business* (St. Louis: R. P. Studley, 1907).
10 Vismann, *Files,* 132. For a detailed history of loose-leafing in American legal offices, see Howard T. Senzel, "Looseleafing the Flow: An Anecdotal History of One Technology for Updating," *American Journal of Legal History* 4, no. 2 (2000): 115–97.
11 Krajewski, *Paper Machines,* 137.
12 Charles W. Wootton and Carel M. Wolk, "The Evolution and Acceptance of the Loose-Leaf Accounting System, 1885–1935," *Technology and Culture* 41, no. 1 (2000): 80–98.
13 Wylie v. Bushnell, 277 Ill. 492 (1917).

14. William D. Seddon, "What Constitutes a Book of Original Entry," *New Jersey Law Journal* 21 (April 1898): 107.
15. Seddon, 107.
16. Lewis v. England, 14 Wyo. 128 (1905); Larue v. Rowland, 7 Barb. 107 (1849).
17. Shepherd v. Butcher Tool & Hardware Co., 198 Ala. 275 (1916).
18. Foothill Ditch Co. v. Wallace Ranch Water Co., 25 Cal. App. 2d 555 (1938).
19. Tabeta v. Murane, 76 Cal. App. 2d 887 (1946).
20. Piper, *Book Was There,* 51.
21. Peter Stallybrass, "Books and Scrolls: Navigating the Bible," in *Books and Readers in Early Modern England: Material Studies,* ed. Jennifer Anderson and Elizabeth Sauer (Philadelphia: University of Pennsylvania Press, 2001), 42–79.
22. Yates, *Control through Communication,* 31–39; Uber, "Creating the Steel Chapel," 104–29.
23. Although neither patents nor the archival materials of office equipment companies acknowledge it, the flat file appears to have a precursor in the "slip box" of the sixteenth century. See Vismann, *Files,* 135.
24. M. J. Wine, *Catalogue of National Office Furniture* (Washington, D.C.: M. J. Wine, 1884), 6, National Museum of American History.
25. Roland W. Butters, "70 Years of Filing Development," *The File,* March 1939, 5, Herkimer County Historical Society.
26. James McCord, *A Textbook of Filing* (New York: D. Appleton, 1920), 2.
27. P. J. Schlicht, Filing Cabinet, U.S. Patent 355,219, filed December 4, 1885, and issued December 28, 1886.
28. Office Specialty Co., *Labor Saving Devices for Mercantile and Public Offices* (n.d.), n.p., National Museum of American History.
29. Yawman and Erbe, *Filing Devices Catalog No. 2816* (Rochester, N.Y.: Yawman and Erbe, 1916), 69, National Museum of American History.
30. Ethel Scholfield, *Filing Department Operation and Control* (New York: Roland Press, 1923), 290.
31. W. D. Wigent, B. D. Housel, and E. H. Gilman, *Modern Filing: A Textbook on Office System* (Rochester, N.Y.: Yawman and Erbe, 1916), 75–77.
32. Wine, *Catalogue of National Office Furniture,* 10.
33. Globe Co., *Catalogue of Office Appliances* (Cincinnati: Globe Co., n.d.), 12, National Museum of American History.
34. For a detailed analysis of the design of these cabinets, see Uber, "Creating the Steel Chapel," 104–29.
35. Globe Co., *Office Appliances: Filing Cabinets* (Cincinnati: Globe Co., 1896), 21, National Museum of American History.
36. Globe Co., 9.
37. Globe Co., *Catalogue of Office Appliances,* 12.
38. C. Harris, Improvement in Drawers for Bureaus, U.S. Patent 199,825, filed December 11, 1877, and issued January 29, 1878.
39. Harris; Joseph W. Dane, Improvement in Drawer-Supports, U.S. Pat-

ent 203,894, filed April 29, 1878, and issued May 21, 1878; Francis Sabot, Drawer-Slide, U.S. Patent 354,744, filed August 30, 1886, and issued December 21, 1886.
40 Wallace Grange, Desk or Cabinet Drawer, U.S. Patent 623,537, filed June 23, 1898, and issued April 25, 1899; Charles Besly and Timothy Sullivan, Drop File Cabinet, U.S. Patent 640,155, filed April 12, 1899, and issued December 26, 1899.
41 Wine, *Catalogue of National Office Furniture*, 7.
42 Schlicht and Field, *Catalogue of Labor Saving Office Devices* (Rochester, N.Y.: Schlicht and Field, n.d.), 26, National Museum of American History.
43 Edward A. Cope, *Filing Systems: Their Principles and Their Application to Modern Office Requirements* (London: Sir Isaac Pitman & Sons, 1913), 18-19.
44 "The Age of Steel," *American Architect and Building News*, January 2, 1904, 83.
45 C. W. Simpson, *How It All Started: Art Metal the First Forty Years* (Jamestown, N.Y.: Art Metal Construction Co., 1928), 17.
46 Art Metal Construction Co., *Steel Office Equipment* (Catalogue 763) (Jamestown, N.Y.: Art Metal Construction Co., 1923), 4, Hagley Museum and Library.
47 Quoted in Allan L. Benson, "The Wonderful New World Ahead of Us: Some Startling Prophecies of the Future as Described by Thomas A. Edison," *Cosmopolitan*, February 1911, 299, 300.
48 "History of GF" (unpublished manuscript, circa 1950); "These Were Not L.B.," *LB File*, May 1921, 25, Herkimer County Historical Society.
49 "Age of Steel," 83.
50 "Increasing Popularity of Steel Furniture," *Office Appliances*, September 1916, 18-19.
51 Republic Steel Corporation, "Your Faithful Servant" (typed radio transcript, 1940), 12, Hagley Museum and Library; Uber, "Creating the Steel Chapel," 143.
52 The following paragraphs on the development of steel draw from Misa, *Nation of Steel*.
53 Uber, "Creating the Steel Chapel," 151.
54 "Metal Sectional Furniture and Looseleaf Records," *Banker's Magazine*, September 1911, 366.
55 Shaw-Walker, *Steel Business Equipment Catalog No. 530* (Muskegon, Mich.: Shaw-Walker, 1931), 5, Hagley Museum and Library.
56 For a useful comparison between the presentation of wooden and metal filing cabinets, see Library Bureau, *Steel and Card Filing Cabinets* (Boston: Library Bureau, 1925), Herkimer County Historical Society; Library Bureau, *Unit Filing Cabinets in Wood* (Boston: Library Bureau, 1925), Herkimer County Historical Society.
57 James Wallen, *Things That Live Forever* (Jamestown, N.Y.: Art Metal Construction Co., 1921), 45.

58 One of the most common strategies involved using five-ply wood instead of solid wood for drawer fronts; this also responded to anxieties about warping. Made with four layers of hardwood and a thicker layer of softwood, these fronts tended to be called "veneers," much to the frustration of manufacturers, who saw the label as an attack on the quality of their products. Other responses to steel, such as removing horizontal frames between drawers, made manufacturing easier (and cheaper), but created the impression of less sturdy cabinets. See Library Bureau, "Salesmen Bulletin 651," August 15, 1922, Herkimer County Historical Society; W. S. Cumming, "Office Furniture," *Office Appliances,* January 1922, 146; Library Bureau, "Salesmen Bulletin 373," September 24, 1914, Herkimer County Historical Society.

59 Library Bureau, "Salesmen Bulletin 373."

60 General Fireproofing, "Who Shall Guard the Guards Themselves?," *Office Appliances,* February 1920, 136.

61 "Metal Sectional Furniture," 366.

62 In fact, the interior of a steel filing cabinet can get so hot that some people have repurposed old metal filing cabinets for smoking meat. Farhan Ahsan, "16 File Cabinet Smokers—Turn an Old File Cabinet into a Meat Smoker," The Self Sufficient Living, October 14, 2017, http://theselfsufficientliving.com.

63 Yawman and Erbe, *Y&E Steel Filing Cabinets, Catalog No. 3801* (Rochester, N.Y.: Yawman and Erbe, 1931), 6, Hagley Museum and Library.

64 Art Metal Construction Co., *Insulated Files* (Jamestown, N.Y.: Art Metal Construction Co., n.d.), F3-2, Fenton Historical Society.

65 Safe-Cabinet Co., *Burned Records* (Marietta, Ohio: Safe-Cabinet Co., 1925), n.p., Hagley Museum and Library.

66 Hugh Wilson Patterson, "How an Art Metal Safe Is Built," *Steel Filings,* February 1918, 4–5, Fenton Historical Society.

67 Safe-Cabinet Co., *The Safe-Cabinet Drawer Safe* (Tonawanda, N.Y.: Rand Kardex Service Corp., n.d.), n.p., Hagley Museum and Library.

68 Safe-Cabinet Co., *Burned Records.*

69 H. H. Suender and J. L. Morgan, *The GF News: 50 Years of Progress* (Youngstown, Ohio: Youngstown Printing, 1952), 23–24.

70 "Burn and Smash in Safety Tests," *New York Times,* December 3, 1922.

71 Van Dorn Iron Works Co., *Steel Office Furniture and Filing Cabinet Supplies* (Cleveland: Van Dorn Iron Works Co., 1920), 3, Hagley Museum and Library; Art Metal Construction Co., *Sectional Counter Filing Equipment* (Jamestown, N.Y.: Art Metal Construction Co., 1928), 3, 18, Fenton Historical Society; "Inside Stuff: Locked Cases Give Maximum Fire Resistance," *Steel Filings,* November 1917, 16, Fenton Historical Society.

72 Shaw-Walker, *Built Like a Skyscraper,* 5.

73 "Shaw-Walker Steel Letter Files," *Office Appliances,* February 1917, 143.

74 Library Bureau, *LB Unit Filing Cabinets in Wood* (Catalogue 708) (Boston: Library Bureau, 1925), 6, Herkimer County Historical Society.

75 Art Metal Construction Co., *Steel Filing Devices: Fixtures and Furniture for Commercial Offices and Insurance Offices* (Jamestown, N.Y.: Art Metal Construction Co., 1912), 3, Fenton Historical Society.
76 Globe-Wernicke, *GW Steel Office Equipment, Catalog No. 825* (Cincinnati: Globe-Wernicke, 1926), 12, Hagley Museum and Library.
77 P. H. Yawman, Drawer Support for Filing Cases, U.S. Patent 883,069, filed March 16, 1905, and issued March 24, 1908, 1.
78 H. J. Huening, Sectional Filing Cabinet, U.S. Patent 757,194, filed December 27, 1902, and issued April 12, 1904.
79 Enoch Ohnstrand, Extension Drawer Slide, U.S. Patent 1,391,004, filed July 21, 1920, and issued September 20, 1921; D. E. Hunter, Cabinet Drawer and Case, U.S. Patent 1,172,711, filed December 26, 1914, and issued February 22, 1916; John Ripson and Erick G. Sampson, Support for Drawers and Cabinets, U.S. Patent 884,525, filed October 26, 1907, and issued April 14, 1908.
80 Steelcase, "Nothing Rolls Like a Ball," NA_02111_006_001, Steelcase Corporate Archives, Grand Rapids, Michigan.
81 Globe-Wernicke, *"Elastic" Cabinets, Catalogue No. 29* (Cincinnati: Globe-Wernicke, 1899), 5, Hagley Museum and Library.
82 Library Bureau, *Unit Cabinets* (Boston: Library Bureau, 1912), 6, National Museum of American History.
83 John Durham Peters, *The Marvelous Clouds: Toward a Philosophy of Elemental Media* (Chicago: University of Chicago Press, 2015), 149.
84 Globe-Wernicke, *"Elastic" Cabinets*, 7.

Chapter 3. Cabinet Logic
1 Leffingwell, *Office Management*, 416.
2 Leffingwell, 415.
3 Leffingwell, 416.
4 Leffingwell, 416.
5 Vismann, *Files*, 98; Mattern, "Intellectual Furnishing." Mattern also uses the phrase "cabinet logic" in the title of a lecture based on her intellectual furnishings article. However, she does not develop cabinet logic as a concept. Shannon Mattern, "Cabinet Logic: A History, Critique, and Consultation on Media Furniture" (IKKM Talk, Bauhaus University, Weimar, Germany, January 20, 2016).
6 Te Heesen, *World in a Box*, 153.
7 Te Heesen, *Newspaper Clipping*, 23; E. R. Hudders, *Indexing and Filing: A Manual of Standard Practice* (New York: Ronald Press, 1919), 67.
8 Indexes were defined as instruments of search, while classification was defined as the grouping of similar materials. The authors of how-to office books discussed indexing in reference to card indexes and classification in reference to subject files. In the teaching of filing, the action of a clerk reading a letter to determine where to file it tended to be referred to as "indexing." In promoting preprinted guide cards, office equipment

companies used "index" to label the system of cards. Hudders, *Indexing and Filing*; Scholfield, *Filing Department Operation*; David Duffield, *Progressive Indexing and Filing for Schools* (Tonawanda, N.Y.: Rand Kardex Bureau for the Library Bureau, 1926). For a useful overview of this literature in the context of the development of record management and for a brief overview of definitions of classification and indexing, see Carol E. B. Choksy, *Domesticating Information: Managing Documents inside the Organization* (Lanham, Md.: Scarecrow Press, 2006), 18–25, 189.

9 Te Heesen, *World in a Box*, 152.
10 Te Heesen, 153.
11 Galloway, *Office Management*, 91.
12 Walker, *Office and Tomorrow's Business*, 125.
13 Csiszar: "Bibliography as Anthropometry."
14 Library Bureau, "Boston Salesmen's Periodical Meetings," 203–4.
15 C. M. Carnahan, Expansible Envelop for Vertical Files, U.S. Patent 847,648, filed January 27, 1906, and issued March 19, 1907.
16 Wigent et al., *Modern Filing*, 12; Cope, *Filing Systems*, 25–26.
17 T. J. Amberg and W. D. Patterson, Filing Case, U.S. Patent 829,750, filed April 4, 1905, and issued August 28, 1906.
18 Hudders, *Indexing and Filing*, 59; A. T. Weiss, Follower for Paper-Filing Devices, U.S. Patent 879,343, filed June 20, 1907, and issued February 18, 1908.
19 "History of Vertical Filing," 48; Hunter, *Modern Filing Manual*, 6.
20 Weiss, Follower for Paper-Filing Devices.
21 Yawman and Erbe, *Y&E Record Filing Cabinets, General Catalog 916e* (Rochester, N.Y.: Yawman and Erbe, 1910), 13, Hagley Museum and Library.
22 Flexifile Co., *Filing Devices* (Chicago: Flexifile Co., n.d.), 2, Hagley Museum and Library.
23 E. G. Sampson, File Holder, U.S. Patent 924,690, filed January 2, 1909, and issued June 15, 1909; Art Metal Construction Co., *User's Guide to Office Modernization: Steel Office Equipment* (Jamestown, N.Y.: Art Metal Construction Co., 1940), 15, Fenton Historical Society; Metal Office Furniture Co., *Steelcase 5-Drawer Letter Files* (Grand Rapids, Mich.: Metal Office Furniture Co., n.d.), Steelcase Corporate Archives, Grand Rapids, Michigan.
24 E. G. Sampson, Filing-Case, U.S. Patent 863,944, filed April 13, 1907, and issued August 20, 1907; D. E. Hunter, Filing Cabinet, U.S. Patent 1,289,180, filed July 14, 1917, and issued December 31, 1918.
25 Art Metal Construction Co., *Vertical Filing Equipment Catalog 766-1* (Jamestown, N.Y.: Art Metal Construction Co., 1928), 4, Fenton Historical Society.
26 Shaw-Walker, "Build Your Reputation-Wall Solid," *Office Appliances*, December 1920, 146.
27 Globe-Wernicke, *Steel Office Equipment Catalog No. 1441* (Cincinnati: Globe-Wernicke, 1941), 7, Furniture Trade Catalogs, Collection 232, Grand Rapids Public Library, Grand Rapids, Michigan (hereafter Grand Rapids Public Library).

28 Metal Office Furniture Co., *Steelcase Stylefiles* (Grand Rapids, Mich.: Metal Office Furniture Co., n.d.), 12, Steelcase Corporate Archives, Grand Rapids, Michigan; Remington Rand, *7 Minutes with the Aristocrat* (Tonawanda, N.Y.: Remington Rand, n.d.), 7–8, Herkimer County Historical Society.
29 Shaw-Walker, *Steel Business Equipment Catalog No. 530*, 9. See also Shaw-Walker, *Office Guide 1940: Modern Equipment and Its Efficient Use* (Muskegon, Mich.: Shaw-Walker, 1940), 33, Grand Rapids Public Library.
30 Art Metal Construction Co., "And We Took the Clutch Out of Filing," *Office Economist*, July–August 1941, n.p., Fenton Historical Society.
31 Globe-Wernicke, *Steel Office Equipment Catalog No. 1441*, 7.
32 American Institute of Filing, *Course in Correspondence Filing*, 36.
33 Shaw-Walker, *The Shaw-Walker Book of System Supplies, Catalog No. 518-E* (Muskegon, Mich.: Shaw-Walker, n.d.), 32, Hagley Museum and Library.
34 Rand Company, *Filing! Old Fashioned or Modern* (North Tonawanda, N.Y.: Rand Company, n.d.), n.p., Hagley Museum and Library.
35 William Wigent, *Teacher's Key: For Use with "Modern Filing: A Textbook on Office System"* (Rochester, N.Y.: Yawman and Erbe, 1917), 20.
36 Shaw-Walker, *The "Ideal" System of Filing Letters (and Finding Them)* (Muskegon, Mich.: Shaw-Walker, n.d.), n.p., Hagley Museum and Library.
37 W. B. Mehl, Catalogue Card, U.S. Patent 968,065, filed August 26, 1908, and issued August 23, 1910.
38 G. H. Dawson, Index Tab, U.S. Patent 1,511,268, filed February 21, 1923, and issued October 14, 1924.
39 Dawson.
40 Globe-Wernicke, "Look at the Guide—Not for It," *The File*, March 1938, 7.
41 Globe-Wernicke, 7.
42 Rand Company, *Filing!*, n.p.
43 Yawman and Erbe, *Vertical Filing Down-to-Date*, 7th ed. (Rochester, N.Y.: Yawman and Erbe, 1941), 13, National Museum of American History.
44 Globe-Wernicke, *Steel Office Equipment Catalog No. 1441*, 7.
45 Globe-Wernicke, 6.
46 Heather Wolfe, "Filing, Seventeenth-Century Style," The Collation (blog), Folger Shakespeare Library, March 28, 2013, https://collation.folger.edu; Craig Robertson, "Files," in *Information: A Historical Companion*, ed. Ann Blair, Paul Duguid, Anja-Silvia Goeing, and Anthony Grafton (Princeton, N.J.: Princeton University Press, 2021).
47 Iana Feldman, *Governing Gaza: Bureaucracy, Authority, and the Work of Rule, 1917–1967* (Durham, N.C.: Duke University Press, 2008); Matthew Hull, *Government through Paper: The Materiality of Bureaucracy in Urban Pakistan* (Berkeley: University of California Press, 2012); Latour, *Making of Law*; Vismann, *Files*.
48 Warwick Anderson, "The Case of the Archive," *Critical Inquiry* 39, no. 3 (Spring 2013): 532–47.
49 On "co-documentary context," see Matthew Hull, "The File: Agency,

Authority, and Autography in an Islamabad Bureaucracy," *Language & Communication* 23 (2003): 295–96.
50 This is a critique of Max Weber's argument that writing exists in the service of organizational control (JoAnne Yates's argument about "communication through control" aligns with this Weberian position). See Hull, *Government through Paper*, 112; Yates, *Control through Communication*.
51 H. S. Brown, *Filing: Theory and Practice* (New York: Hubbard, 1933), 8.
52 Library Bureau, "Boston Salesmen's Periodical Meetings," 90.
53 Wigent et al., *Modern Filing*, 13. For additional examples, see Margaret A. Lennig, *Filing Methods: A Textbook on the Filing and Indexing of Commercial and Government Records* (Philadelphia: T. C. Davis & Sons, 1920), 21; McCord, *Textbook of Filing*, 20; Hudders, *Indexing and Filing*, 58.
54 John L. Stephens, "Standardization in the Filing Equipment Industry," *Annals of the American Academy of Political and Social Science* 137 (May 1928): 184; A. D. Dunn, *Notes on the Standardization of Paper Sizes* (London, Ontario: London Free Press, 1973), 6.
55 Nader Vossoughian, "Standardization Reconsidered: *Normierung* in and after Ernst Neufert's *Bauentwurfslehre* (1936)," *Grey Room*, no. 54 (Winter 2014): 39.
56 Scholfield, *Filing Department Operation*, 267.
57 Edna Cotta, "The Organization and Management of a Centralized File Bureau," *The File*, September 1935, 6–7, Herkimer County Historical Society.
58 Irene Julian, "Filing in a Law Office," *Filing*, September 1918, 84.
59 Ellen L. Eastman, "Filing for Law Offices," *Office Economist*, July–August 1934, 5–7, 12–13; McCord, *Textbook of Filing*, 107–8; Scholfield, *Filing Department Operation*, 228–29.
60 John M. Hollingsworth and Lyman Hollingsworth, Improvement in Paper Making, U.S. Patent 3,362, December 4, 1843.
61 Leffingwell, *Office Management*, 425–26.
62 Sulfate folders were made from pulp produced from wood chips cooked under pressure in a solution of caustic soda. No ground wood or other adulterant was present. This made for strong fold and tear resistance, qualities that must be present if a folder is to withstand hard and long wear. R. G. Beal, "The Selection and Use of Filing Supplies," *Office Economist*, November–December 1949, 14, Fenton Historical Society.
63 Isaac Wagemaker, *A Comprehensive Textbook on Business and Office Systematizing, Vertical Filing Systems, Filing Devices, Card Ledger System, Profit Showing System for a Wholesale or Manufacturing Business etc.*, vol. 1 (Grand Rapids, Mich.: Isaac Wagemaker, 1907), 62.
64 American Institute of Filing, *Course in Correspondence Filing*, 39.
65 Wigent, *Teacher's Key*, 14.
66 "What Do You Know about Filing? To Properly and Efficiently Organize You Must Be Better Versed in Work That Is Performed Than Those Doing It," *Office Economist*, September 1919, 117, Fenton Historical Society.

67 For a critique of vertical filing based on the space that folders took up in a file drawer, see Elbert D. Murphy, "The Way of the File," *Filing,* September 1919, 457–58.
68 American Institute of Filing, *Course in Correspondence Filing,* 38–39.
69 Remington Rand, *The Office Manual of Better Filing Supplies* (Buffalo, N.Y.: Remington Rand, n.d.), 13, National Museum of American History.
70 Jon Agar, *The Government Machine: A Revolutionary History of the Computer* (Cambridge, Mass.: MIT Press, 2003), 1–2.
71 Agar, 3.
72 Nunberg, "Farewell to the Information Age," 116–17.

Chapter 4. Granular Certainty

1 G. Lynn Sumner, *How I Learned the Secrets of Success in Advertising* (1952; repr., n.p.: North Audley Media, 2009), 25.
2 "Indexing and Recording of Patterns," *System,* December 1901, n.p.
3 This is a critical-historical take on James Beniger's argument that "information . . . is an entity that becomes salient in a certain kind of society—a society of great speed, complexity, and size." Peters, "Control of Information," 6; Beniger, *Control Revolution.*
4 Von Oertzen, "Machineries of Data Power"; Müller-Wille, "Names and Numbers," 109–28.
5 George Frederick, "More Time to Sell," *System,* February 1914, 184; Ethel Erickson, *The Employment of Women in Offices* (Bulletin of the Women's Bureau, no. 120) (Washington, D.C.: Government Printing Office, 1934), 109; Lars Heide, "Scale and Scope in American Key-Set Office Machine Dynamics, 1880s–1930s," *IEEE Annals of the History of Computing* 33, no. 3 (July–September 2011): 22–31.
6 Chandler, *Visible Hand.*
7 Caitlin Rosenthal, *Accounting for Slavery: Masters and Management* (Cambridge, Mass.: Harvard University Press, 2018), 85–120.
8 Yates, *Control through Communication,* 3; Tamara Plakins Thornton, *Nathaniel Bowditch and the Power of Numbers: How a Nineteenth-Century Man of Business, Science, and the Sea Changed American Life* (Chapel Hill: University of North Carolina Press, 2016); Melissa Gregg, *Counterproductive: Time Management in the Knowledge Economy* (Durham, N.C.: Duke University Press, 2018), 33, 49.
9 Joseph Litterer, "Alexander Hamilton Church and the Development of Modern Management," *Business History Review* 35, no. 2 (1961): 211–25; Joseph Litterer, "Systematic Management: Design for Organizational Recoupling in American Manufacturing Firms," *Business History Review* 37, no. 4 (1963): 369–91; Joseph Litterer, "Systematic Management: The Search for Order and Integration," *Business History Review* 35, no. 4 (1961): 461–76.
10 Noble, *America by Design,* 263.
11 Noble, 310.

12 This critique is absent from the celebrated but functionalist writings of Alfred Chandler. See Philip Scranton, "Beyond Chandler?," *Enterprise & Society* 9, no. 3 (2008): 426–29; Evgeny Morozov, "Capitalism's New Clothes," *The Baffler*, February 4, 2019, https://thebaffler.com.
13 Shenhav, *Manufacturing Rationality*, 188.
14 Shenhav, 188.
15 Edward L. Wedeles, "How I Control Expense," *System*, February 1910, 158.
16 John H. McDonald, "Records as an Aid in Managerial Control," *Office Economist*, January 1925, 3, Fenton Historical Society.
17 W. H. Leffingwell, *The First Half Century of Office Management* (New York: National Office Management Association, 1930), 9; W. H. Leffingwell, *Fundamentals of Scientific Office Management* (London: Management Research Groups, 1929), 19. On the impact of scientific management on the office, see Ingrid Jeacle and Lee Parker, "The 'Problem' of the Office: Scientific Management, Governmentality and the Strategy of Efficiency," *Business History* 55, no. 7 (2013): 1074–99.
18 Meister, "From Form to Norm," 69.
19 Meister, 63.
20 JoAnne Yates, "The Emergence of the Memo as a Managerial Genre," *Management Communication Quarterly* 2, no. 4 (1989): 496–506. For a different take on the memo (but one that is still sympathetic to Yates's viewpoint), see John Guillory, "The Memo and Modernity," *Critical Inquiry* 31, no. 1 (Autumn 2004): 108–32.
21 Quoted in Yates, *Control through Communication*, 86.
22 Yates, 80–85.
23 Yates, xvi.
24 Yates, 15; JoAnne Yates, "Evolving Information Use in Firms, 1850–1920: Ideology and Information Techniques and Technologies," in *Information Acumen: The Understanding and Use of Knowledge in Modern Business*, ed. Lisa Bud-Frierman (London: Routledge, 1994), 30.
25 Benge, *Cutting Clerical Costs*, 59.
26 Walker, *Office and Tomorrow's Business*, 96.
27 Haigh, "Technology, Information, and Power," 58–94.
28 Agar, *Government Machine*, 144.
29 Delphine Gardey, "Culture of Gender, and Culture of Technology: The Gendering of Things in France's Office Spaces between 1890 and 1930," in *Cultures of Technology and the Quest for Innovation*, ed. Helga Nowotny (New York: Berghahn Books, 2006), 81.
30 Cotta, "Organization and Management of a Centralized File Bureau," 1.
31 Cotta, 7.
32 Edith McWilliams, "Our Visit to the Brooklyn Edison," *The File*, February 1938, 4, Herkimer County Historical Society.
33 McWilliams, 4.
34 McWilliams, 5.
35 Norma Louise Cohen, "Suggestions for Improving Teaching Procedures

and Devices in the Teaching of Filing," *The File,* May 1938, 5, Herkimer County Historical Society; Remington Rand, *Variadex: The Speed Index with Unlimited Expansion* (Buffalo, N.Y.: Remington Rand, 1944), 32, National Museum of American History.

36 Arthur Frey, "Filing as a Profession," *The File,* June 1939, 1, Herkimer County Historical Society.
37 "Link Belt Company's LB Stock Record," *LB File,* February 1921, 3–5, Herkimer County Historical Society.
38 Gertrude Edelman, "Medical Records," *The File,* October 1935, 4, Herkimer County Historical Society.
39 "Hospital Records," *LB File,* April 1923, 6, Herkimer County Historical Society.
40 Ellen G. Cardwell, "Filing in the Record Room of the Bryn Mawr Hospital," *The File,* February 1935, 6, Herkimer County Historical Society.
41 Edelman, "Medical Records," 1, 4.
42 Rand Company, *Sales Control* (North Tonawanda, N.Y.: Rand Company 1919), 7, Hagley Museum and Library.
43 Rand Company, *Visible Business Control: A Booklet for Executives* (North Tonawanda, N.Y.: Rand Company, n.d.), Hagley Museum and Library; Acme Card Systems Co., *Profitable Business Control with Acme Visible Records* (Chicago: Acme Card Systems Co., n.d.), Hagley Museum and Library; Globe-Wernicke, *Modern Business Control* (Cincinnati: Globe-Wernicke, 1933), National Museum of American History.
44 The description of the commission's work is quoted in Bess Glenn, "The Taft Commission and the Government's Record Practices," *American Archivist* 21, no. 3 (1958): 277. Regarding records management, see Maynard Brichford, "The Relationship of Records Management Activities to the Field of Business History," *Business History Review* 46, no. 2 (1972): 220.
45 Glenn, "Taft Commission," 291.
46 Glenn, 290. The Department of the Interior and the Department of Justice subsequently employed the same company.
47 Books on indexing and filing frequently took from library literature a definition of index as an "instrument of search for specific information." Hudders, *Indexing and Filing,* 9.
48 Scholfield, *Filing Department Operation,* 277.
49 "American Mills Co.," *LB File,* December 1921, 15, Herkimer County Historical Society.
50 "Belknap Hardware Company: File Installation," *LB File,* December 1921, 3–4, Herkimer County Historical Society.
51 "Direct Alphabetic Files for Bond House," *LB File,* January 1923, 8, Herkimer County Historical Society.
52 Helen Bryant, "*Life*'s Picture Library," *The File,* November 1939, 6–7, Herkimer County Historical Society; "Movie Morgue," *LB File,* April 1923, 9, Herkimer County Historical Society.
53 For examples of different techniques for filing negatives in personal

photograph collections, see Paul L. Miller, "A Film Negative File and How to Make One," *Photo Era,* November 1928, 247–53; Fred E. Kunkel, "A World of Ideas at One's Finger-Tips," *Photo Era,* May 1926, 242–44; R. C. Pepper, "Filing, Storing and Indexing Miniature Negatives," *American Photography,* October 1934, 646–48.

54 "Filing Newspaper Clippings," *Filing,* November–December 1918, 16.
55 Te Heesen, *Newspaper Clipping,* 10.
56 Margaret A. McVey and Mabel E. Colgrove, "Section 1: The Vertical File," in *Modern American Library Economy as Illustrated by the Newark N.J. Free Public Library,* vol. 2, pt. 18, by John Cotton Dana (Woodstock, Vt.: Elm Tree Press, 1915), 104.
57 McVey and Colgrove, 101; Della G. Ovitz and Zana K. Miller, *A Vertical File for Every Library* (Buffalo, N.Y.: Library Bureau Division, Remington Rand, n.d.), 1, Herkimer County Historical Society.
58 McVey and Colgrove, "Section 1," 104.
59 McVey and Colgrove, 117; Ovitz and Miller, *Vertical File for Every Library,* 7.
60 Library Bureau, "Salesmen Bulletin 449," February 14, 1919, 612, Herkimer County Historical Society.
61 Shaw-Walker, *Office Guide 1940,* 50.
62 "Home Savings Bank, Albany: Uses L.B. Ledger and Signature Cards," *LB File,* April 1923, 4, Herkimer County Historical Society; "Bank Department," *LB File,* November 1921, 12, Herkimer County Historical Society; Wigent et al., *Modern Filing,* 77–79.
63 Hiram Blauvelt, "Filing Blue Prints," *System,* April 1927, 512.
64 Erich Schmied, "Filing Blueprints," *Architectural Record,* May 1931, 425–26.
65 Bertha M. Weeks, "Why Engineering Drawings Demand a Filing System of Their Own," *Office Economist,* September–October 1938, 6, Fenton Historical Society.
66 Hamilton Manufacturing Company, *Drafting Room Furniture* (Two Rivers, Wis.: Hamilton Manufacturing Company, 1936), 6, National Museum of American History.
67 Hamilton Manufacturing Company, 13.
68 Art Metal Construction Co., *Art Metal Planfiling* (Jamestown, N.Y.: Art Metal Construction Co., 1928), 4, Hagley Museum and Library.
69 Yawman and Erbe, *Steel Filing Equipment, Catalog No. 3500* (Rochester, N.Y.: Yawman and Erbe, 1922), 44, Hagley Museum and Library; American Institute of Filing, *Course in Correspondence Filing,* 164, Herkimer County Historical Society.
70 Globe-Wernicke, *Cello-Clip Map and Plan File* (Cincinnati: Globe-Wernicke, 1935), Grand Rapids Public Library.
71 W. C. Reavis and Robert Woellnor, "Labor-Saving Devices Used in Office Administration in Secondary Schools," *School Review* 36, no. 10 (December 1928): 737–38.
72 Reavis and Woellnor, 743, 744.

73 Donald Bond, "A Method of Teaching Contemporary Poetry," *English Journal* 12, no. 10 (December 1923): 679–85. See also Roger W. Boop, "Resources at Your Finger Tips," *Clearing House* 39, no. 5 (January 1965): 283–85.
74 Francis Stuart Chapin, "Business System in the Professor's Study," *School and Society* 2 (1915): 710.
75 "Sermon File," *LB File*, November 1922, 18, Herkimer County Historical Society.
76 McCord, *Textbook of Filing*, vi.
77 "Indexing and Recording of Patterns."
78 Thomas Osborne, "The Ordinariness of the Archive," *History of the Human Sciences* 12, no. 2 (1999): 58.
79 John F. Arndt, "Making a File Drawer Live," *The File*, February 1936, 10, Herkimer County Historical Society.
80 Leffingwell, *Office Management*, 831.
81 Stephens, "Standardization in the Filing Equipment Industry," 182. See also Hunter, *Modern Filing Manual*, xv.
82 Walter A. Friedman, *Fortune Tellers: The Story of America's First Economic Forecasters* (Princeton, N.J.: Princeton University Press, 2013).
83 William Henry Leffingwell, ed., *The Office Appliance Manual* (New York: National Association of Office Appliance Manufacturers, 1926), 6.

Chapter 5. Automatic Filing

1 Impressions, "The Filing Cabinet," *Steel Filings*, June 1918, 15, Fenton Historical Society.
2 For an analysis of the metaphor of the machine, see Agar, *Government Machine*, 15–44.
3 *OED Online*, s.v. "machine," accessed September 14, 2020, http://www.oed.com.
4 Lisa Gitelman, *Scripts, Grooves, and Writing Machines: Representing Technology in the Edison Era* (Stanford, Calif.: University of Stanford Press, 1999), 189.
5 Eugenia Wallace, *Filing Methods* (New York: Ronald Press, 1924), 72.
6 J. A. Cramer, *The Filing Department* (New York: Bankers Publishing, 1917), 27.
7 Mattern, "Intellectual Furnishing."
8 Shaw-Walker, *Shaw-Walker Book of System Supplies*, 42.
9 G. C. Kingsley, "The L.B. Automatic Index," *LB File*, July–September 1923, 3, Herkimer County Historical Society.
10 Library Bureau, *Filing as a Profession for Women* (Boston: Library Bureau, 1919), 31–32.
11 McCord, *Textbook of Filing*, 28–29; Hudders, *Indexing and Filing*, 60–61.
12 Fred E. Kunkel, "Routing Sheet or Rubber Stamp: How Routing Systems Speed Up and Check the Daily Mail," *Filing*, November 1920, 675, 681.
13 Cope, *Filing Systems*, 17.
14 Brown, *Filing*, 1.

15 George Darlington, *Office Management* (New York: Ronald Press, 1935), 25, 141.
16 Scholfield, *Filing Department Operation*, 3–4.
17 Csiszar, "Bibliography as Anthropometry," 442–43.
18 Scholfield, *Filing Department Operation*, 118, 3.
19 Scholfield, 3, 97. Two decades after Scholfield, Vannevar Bush proposed a machine he called the Memex, which is now viewed as anticipating hypertext. In explaining the Memex, Bush argued that the storage of information should be modeled on human consciousness. However, he saw little value in "bureaucratic indexing systems." For him the mechanical associations of the filing cabinet that Scholfield saw as cause for celebration were artificial. People did not think in alphabetical or numerical order. Rather than the rigidity of these hierarchical systems, Bush advocated "selection by association." This process would emulate the "web of trails" that connect a person's thoughts. Vannevar Bush, "As We May Think," in *From Memex to Hypertext: Vannevar Bush and the Mind's Machine,* ed. James M. Nyce and Paul Kahn (Boston: Academic Press, 1991), 88–109; Theodore H. Nelson, "As We Will Think," in Nyce and Kahn, *From Memex to Hypertext*, 245–60.
20 Gitelman, *Scripts, Grooves, and Writing Machines*, 189.
21 Matt K. Matsuda, *The Memory of the Modern* (New York: Oxford University Press, 1996), 92.
22 Flexifile, *Filing Devices*, 23.
23 Mary Carruthers, *The Book of Memory: A Study in Medieval Culture,* 2nd ed. (Cambridge: Cambridge University Press, 2008), 37–55.
24 Carruthers, 37–38.
25 Carruthers, 22–23.
26 Francis A. Yates, *The Art of Memory* (Chicago: University of Chicago Press, 1966). For a critique of historians' reliance on Yates's work and place-based metaphors in discussions of memory, see Eric Garberson, "Libraries, Memory, and the Space of Knowledge," *Journal of the History of Collections* 18, no. 2 (2006): 112–16.
27 Carruthers, *Book of Memory*, 38.
28 "Man with the Camera Eyes," *Gulfport Daily Herald,* January 24, 1910, 3. See also "Mr. Duke, Genius of Construction; Something of the Personal Side of Southern Power Company Head," *Charlotte Daily Observer,* November 29, 1915, 6.
29 The sixteenth-century development of floor-to-ceiling bookshelves offers another example of how the vertical storage of paper (in books, standing on shelves) allowed a person to survey the contents of a collection. However, in contrast to ideas of efficiency and system, explanations for this development invoked the older arts of memory and newer discourses about encyclopedic order and learning. See Garberson, "Libraries, Memory, and the Space of Knowledge," 105.
30 Library Bureau, *A Few Letters from Users of the Automatic Index* (Boston: Library Bureau, 1913), n.p., Herkimer County Historical Society.

31 Yawman and Erbe, *Vertical Filing Down-to-Date,* 13.
32 Alberto Cevolini, "Storing Expansions: Openness and Closure in Secondary Memories," in *Forgetting Machines: Knowledge Management Evolution in Early Modern Europe,* ed. Alberto Cevolini (Leiden, Netherlands: Brill, 2016), 155–87. Cevolini uses the problematic and anachronistic name "filing cabinet" to refer to the slipcases and different cabinets that were designed to hold the cards that began to be used in note-taking during the early modern period.
33 These books were usually organized according to argument or narrative, which made the material more amenable to the utility that cross-referencing promised. In contrast, the books that filing cabinets and loose-leaf binders replaced were organized chronologically.
34 For an insightful analysis of index card systems and cross-referencing, see Markus Krajewski, "Paper as Passion: Niklas Luhmann and His Card Index," trans. Charles Marcum II, in *Raw Data Is an Oxymoron,* ed. Lisa Gitelman (Cambridge, Mass.: MIT Press, 2013), 103–20.
35 Cevolini, "Storing Expansions," 173; Alberto Cevolini, "Knowledge Management Evolution in Early Modern Europe: An Introduction," in Cevolini, *Forgetting Machines,* 20. For the Luhmann quote, see Krajewski, "Paper as Passion," 113.
36 Krajewski, *Paper Machines,* 7.
37 A filing cabinet could incorporate cross-referencing, but it was not primarily designed for such a system. When cross-referencing was used with a filing cabinet, it was usually a subject system that took the form of cards with "see also" notations rather than a separate index of keywords.
38 Globe-Wernicke, "He's Our New File Clerk," advertising blotter, n.d., author collection; Globe-Wernicke, "You Had Better Get New Files or a Memory Course for Me!," *The File,* September 1939, 7, Herkimer County Historical Society; Dickson School of Memory, "Stop Forgetting," *System,* December 1911, advertising section, n.p.
39 Alan Liu, "Transcendental Data: Toward a Cultural History and Aesthetics of the New Encoded Discourse," *Critical Inquiry* 31, no. 1 (2004): 72.
40 John Roberts, *The Intangibilities of Form: Skill and Deskilling in Art after the Readymade* (London: Verso, 2007), 83.
41 Carla Bittel, Elaine Leong, and Christine von Oertzen, "Paper, Gender, and the History of Knowledge," in *Working with Paper: Gendered Practices in the History of Knowledge,* ed. Carla Bittel, Elaine Leong, and Christine von Oertzen (Pittsburgh: University of Pittsburgh Press, 2019), 9.
42 John Durham Peters, "Writing," in *The International Encyclopedia of Media Studies,* vol. 5, *Media Effects/Media Psychology,* ed. Angharad N. Valdivia and Erica Scharrer (Malden, Mass.: Blackwell, 2013), 205.
43 Friedrich Kittler, *Gramophone, Film, Typewriter* (Stanford, Calif.: Stanford University Press, 1999), 183–265.
44 Gitelman, *Scripts, Grooves, and Writing Machines,* 188.
45 Sharon Hartman Strom, *Beyond the Typewriter: Gender, Class, and the Origins*

of Modern American Office Work, 1900–1930* (Urbana: University of Illinois Press, 1992), 177.
46 Impressions, "Filing Cabinet," 15.
47 McCord, *Textbook of Filing*, 19.
48 Tyrus Miller, "Rethinking the Digital Hand," October 18, 2013, https://crosspollenblog.wordpress.com; Jacques Derrida, "Heidegger's Hand (Geschlecht II)," in *Deconstruction and Philosophy: The Texts of Jacques Derrida*, ed. John Sallis (Chicago: University of Chicago Press, 1989), 161–96.
49 Brown and Duguid, *Social Life of Information*, 119–20.
50 Yawman and Erbe, *The Executive's Workshop: A Booklet on Efficient Office Management* (Rochester, N.Y.: Yawman and Erbe, 1922), 11–12, Hagley Museum and Library.
51 Gardey, "Culture of Gender," 87.
52 Strom, *Beyond the Typewriter*, 83–90, 185.
53 For discussion of a similar relationship between the representation of women and the perception of women as workers in advertisements and work in the context of British computing in the middle of the twentieth century, see Mar Hicks, "Only the Clothes Changed: Women Operators in British Computing and Advertising, 1950–1970," *IEEE Annals of the History of Computing* 32, no. 2 (2010): 2–14.
54 For another example of an advertising campaign using close-ups of hands in the 1920s and 1930s, see Kate Forde, "Celluloid Dreams: The Marketing of Cutex in America, 1916–1935," *Journal of Design History* 15, no. 3 (2002): 175–88.
55 For another example of the representation of labor in the discourse of efficiency that detaches labor from the body, see Sharon Corwin's analysis of Frank and Lillian Gilbreth's time-exposure films that registered workers' labor as lines of light, "Picturing Efficiency: Precisionism, Scientific Management, and the Effacement of Labor," *Representations* 84, no. 1 (2003): 139–48.
56 Roberts, *Intangibilities of Form*, 93.
57 Janet Zandy, *Hands: Physical Labor, Class, and Cultural Work* (New Brunswick, N.J.: Rutgers University Press, 2004), 181.
58 Shaw-Walker, *Office Guide* (Muskegon, Mich.: Shaw-Walker Co., 1949), 74, Grand Rapids Public Library. This description of an index card cabinet illustrates the cabinet logic shared by the index card cabinet and the filing cabinet. For a history of the card catalog that emphasizes its media specificity and distinctiveness from the filing cabinet, see Krajewski, *Paper Machines*.
59 Wallace, *Filing Methods*, 12.
60 Braverman, *Labor and Monopoly Capital*, 220.
61 Rachel Plotnick, *Power Button: A History of the Pleasure, Panic, and Politics of Pushing* (Cambridge, Mass.: MIT Press, 2018), xv.
62 Plotnick, 111.
63 Plotnick, 111.

64 Kenneth Lipartito argues that the job of telephone operator developed with the belief that "manual switching had a gender." Kenneth Lipartito, "When Women Were Switches: Technology, Work, and Gender in the Telephone Industry, 1890–1920," *American Historical Review* 99, no. 4 (1994): 1082, 1084.

65 Quoted in Jennifer Light, "When Computers Were Women," *Technology and Culture* 40, no. 3 (1999): 461.

66 Charles L. Pederson, "Three Million Customers and the Filing Problems They Create," *The File*, February 1936, 6, Herkimer County Historical Society. Piano playing was also used to link women to the typewriter; see Christopher Keep, "The Cultural Work of the Type-Writer Girl," *Victorian Studies* 40, no. 3 (Spring 1997): 405.

67 Judith A. McGaw, *Most Wonderful Machine: Mechanization and Social Change in Berkshire Paper Making, 1801–1885* (Princeton, N.J.: Princeton University Press, 1987), 353.

68 McGaw, 342.

69 McGaw, 338.

70 Stephen Mihm, "Clerks, Classes, and Conflicts: A Response to Michael Zakim's 'The Business Clerk as Social Revolutionary,'" *Journal of the Early Republic* 26, no. 4 (2006): 609.

71 Quoted in Michael Zakim, "Producing Capitalism: The Clerk at Work," in *Capitalism Takes Command*, ed. Michael Zakim and Gary Kornblith (Chicago: University of Chicago Press, 2012), 244.

72 Zakim, *Accounting for Capitalism*, 146, 96.

73 Contrary to this argument, Zakim contends that the arrival of the typewriter and women did not change the office in any fundamental way. Zakim, 31. However, Zakim's belief that the modern office appeared in the nineteenth century causes him to downplay the constitutive effect of the arrival of machines and women as machine operators on the emergence of the modern office in the early twentieth century.

74 King Features Syndicate, "Barney, the File Clerk, Gets Fired. Will He Get the Raise?," *Filing and Office Management*, July 1921, 42, 43.

75 Shaw-Walker, *The Ideal System of Filing Letters (and Finding Them)* (Muskegon, Mich.: Shaw-Walker, n.d.), n.p., Hagley Museum and Library.

76 Shoshana Zuboff, *In the Age of the Smart Machine: The Future of Work and Power* (New York: Basic Books, 1988), 151.

Chapter 6. The Ideal File Clerk

1 Galloway, *Office Management*, 157.

2 For discussion of the similar conflation between women and typewriters, see Thomas Schlereth, *Cultural History and Material Culture: Everyday Life, Landscapes, Museums* (Charlottesville: University Press of Virginia, 1990), 158–59.

3 For a discussion of class, gender, and race in the context of early twentieth-century office work, see Kwolek-Folland, *Engendering Business*, 11–14.

4 Ben Kafka, *The Demon of Writing: Powers and Failures of Paperwork* (New York: Zone Books, 2012).
5 A. M. Edwards, *Sixteenth Census of the United States 1940: Population: Comparative Occupation Statistics for the United States, 1870 to 1940* (Washington, D.C.: Government Printing Office, 1943), 113, 121.
6 Ruth Shonle Cavan, "The Girl Who Writes Your Letters," *Survey*, July 15, 1929, 439.
7 Cavan, 439.
8 Erickson, *Employment of Women in Offices*, 11–12, 14; M. C. Elmer, *A Study of Women in Clerical and Secretarial Work in Minneapolis, Minn.* (Minneapolis: Women's Occupational Bureau, 1925), 29.
9 Elmer, *Study of Women*, 28.
10 While it is possible that for some women the office provided an escape from the boredom of the home, it also seems likely that most women were not truly satisfied doing routine clerical work. This comment is only speculative, however, given that accounts of the experiences of low-level clerical workers are almost completely absent from archival records.
11 Erickson, *Employment of Women in Offices*, 2–3.
12 Charl Williams, "The Position of Women in the Public Schools," *Annals of the American Academy of Political and Social Science* 143 (May 1929): 156.
13 Erickson, *Employment of Women in Offices*, 6.
14 Erickson, 6. File clerks were the fourth-largest occupational group, behind general clerk, stenographer, and machine operator. A smaller 1925 survey of Minneapolis office workers identified 4.6 percent of office workers as file clerks. Elmer, *Study of Women*, 6.
15 Erickson, *Employment of Women in Offices*, 8; Harriet A. Byrne, *Women Who Work in Offices* (Bulletin of the Women's Bureau, no. 132) (Washington, D.C.: Government Printing Office, 1935), 3; Frederick Nichols, *Commercial Education in the High School* (New York: Appleton Century Press, 1933), 318.
16 Erickson, *Employment of Women in Offices*, 10.
17 Julie Berebitsky, *Sex and the Office: A History of Gender, Power, and Desire* (New Haven, Conn.: Yale University Press, 2012), 8, 59.
18 Berebitsky, 21–140; Lisa Fine, *The Souls of the Skyscraper: Female Clerical Workers in Chicago, 1870–1930* (Philadelphia: Temple University Press, 1990), 65–75, 140–51; Keep, "Cultural Work of the Type-Writer Girl," 401–26; Mills, *White Collar*, 200–204.
19 Fine, *Souls of the Skyscraper*, 143–44.
20 Better Service, "The File Girl," *Filing*, July 1918, 19.
21 Victoria Olwell, "The Body Types: Corporeal Documents and Body Politics, circa 1900," in *Literary Secretaries/Secretarial Culture*, ed. Leah Price and Pamela Thurschwell (Aldershot, England: Ashgate, 2005), 50.
22 Kafka, *Demon of Writing*, 14.
23 Elizabeth King McDowall, "The Requisites of a Good File Clerk," *Filing*, March 1921, 760.

24 Cavan, "Girl Who Writes Your Letters," 438; Ruth Shonle Cavan, *Business Girls: A Study of Their Interests and Problems* (Chicago: Religious Education Association, 1929), 48.
25 Cavan, "Girl Who Writes Your Letters," 438.
26 Erickson, *Employment of Women in Offices,* 14.
27 Strom, *Beyond the Typewriter,* 190–96, 388–94.
28 Mar Hicks, *Programmed Inequality: How Britain Discarded Women Technologists and Lost Its Edge in Computing* (Cambridge, Mass.: MIT Press, 2017), 234.
29 Stuart M. Blumin, "The Hypothesis of Middle-Class Formation in Nineteenth-Century America: A Critique and Some Proposals," *American Historical Review* 90, no. 2 (1985): 299–338; Stuart M. Blumin, *The Emergence of the Middle Class: Social Experience in the American City, 1760–1900* (Cambridge: Cambridge University Press, 1989), 258–97.
30 Schlereth, *Cultural History and Material Culture,* 158.
31 McDowall, "Requisites of a Good File Clerk," 758.
32 This connection was pointed out to me by an anonymous reviewer of my contribution to Sharma and Singh, *MsUnderstanding Media.*
33 Alan Liu makes a similar argument about late twentieth-century knowledge workers. See Alan Liu, *The Laws of Cool: Knowledge Work and Culture* (Chicago: University of Chicago Press, 2004), 123.
34 Frances Maule, "Women Are So Personal," *Independent Women,* September 1934, 280, 301.
35 McDowall, "Requisites of a Good File Clerk," 760; Schlereth, *Cultural History and Material Culture,* 158.
36 Cavan, *Business Girls,* 72–73.
37 Judith A. McGaw, "Why Feminine Technologies Matter," in *Gender and Technology: A Reader,* ed. Lee Wilson (Baltimore: Johns Hopkins University Press, 2003), 27–28.
38 Quoted in Strom, *Beyond the Typewriter,* 371.
39 Strom, 373.
40 Maule, "Women Are So Personal," 280; Strom, *Beyond the Typewriter,* 391.
41 Anita Rapone, "Clerical Labor Force Formation: The Office Woman in Albany, 1870–1930" (PhD diss., New York University, 1981), 198–99.
42 Olivier Zunz, *Making America Corporate, 1870–1920* (Chicago: University of Chicago Press, 1992), 117.
43 Anne Petersen, "Office Girls Get Personality Tips," *The File,* January 1938, 4, Herkimer County Historical Society.
44 Library Bureau, *Filing as a Profession for Women,* 45.
45 "Beauty Chats: The Decline of the Corset," *Charlotte News,* August 20, 1922, 15.
46 Stenographers, bookkeepers, ledger clerks, typists, cashiers, and dictating machine operators were expected to have a high school education. Nichols, *Commercial Education in the High School,* 316.
47 Strom, *Beyond the Typewriter,* 280; Janice Weiss, "A History of Commercial

Education in the United States since 1850" (PhD diss., Harvard University, 1978), 204.
48. Ruth Cavan, "Education for Business Girls," *Survey* 62 (1929): 601–2.
49. Weiss, "History of Commercial Education," 95–120; Leverett S. Lyon, *Education for Business* (Chicago: University of Chicago Press, 1922), 408–10.
50. Quoted in Strom, *Beyond the Typewriter*, 279.
51. Samuel Haber, *Efficiency and Uplift: Scientific Management in the Progressive Era, 1890–1920* (Chicago: University of Chicago Press, 1964), ix.
52. Weiss, "History of Commercial Education," 171–76.
53. Wigent, *Teacher's Key*, 5.
54. Library Bureau, "Salesmen's Help No. 17: L.B. Practice Method of Teaching," October 1, 1923, 9–10, 16–18, Herkimer County Historical Society.
55. O. E. Beach, letter to D. W. Duffield, November 6, 1923, Herkimer County Historical Society.
56. David Duffield, "Your Request for More Information," n.d., Herkimer County Historical Society; B. H. Parker, letter to J. L. Rowley, August 6, 1925, Herkimer County Historical Society.
57. Library Bureau, "Salesmen's Help," 7; Library Bureau, *L.B. Practice Method of Teaching* (Boston: Library Bureau), 2, Herkimer County Historical Society.
58. Wigent, *Teacher's Key*, 5.
59. Library Bureau, "Salesmen's Help," 18.
60. Brown, *Filing*, 1.
61. American Institute of Filing, *Course in Correspondence Filing*, 54.
62. American Institute of Filing, 54. A similar strategy was applied in the listening practices of telephone operators. Michèle Martin describes a policy she calls "civil listening." This created the expectation that a woman "ought to attempt to hear only the 'sounds' of callers' voices and to ignore their meaning." Michèle Martin, *"Hello, Central?": Gender, Technology, and Culture in the Formation of Telephone Systems* (Montreal: McGill-Queens University Press, 1991), 69.
63. Guy E. Marion, "The Growing Importance of Subject Filing," *Filing*, December 1919, 508.
64. Marion, 509.
65. American Institute of Filing, *Course in Correspondence Filing*, 223–24.
66. Library Bureau, "Salesmen's Help," 9.
67. Accounting presents an example of another (more successful) attempt to gain professional status for an occupation that involved negotiations about the status of machine work (bookkeeping) and a gendered division of labor. Strom, *Beyond the Typewriter*, 83–90.
68. Library Bureau, *Filing as a Profession for Women*.
69. "University of Vermont Offers Course in Filing," *LB File*, August 1921, 15, Herkimer County Historical Society.
70. Library Bureau, *Filing as a Profession for Women*, 21, 23.
71. Library Bureau, 19.

72 Vance Thompson, *Woman* (New York: E. P. Dutton, 1917), 28.
73 Quoted in Library Bureau, *Filing as a Profession for Women,* 19. Christine von Oertzen shows how orderliness, defined as one of the core values of nineteenth-century housewifery, was critical to the employment of middle-class women in their homes to tabulate and compile data from the Prussian census at the end of the nineteenth century. Christine von Oertzen, "Keeping Prussia's House in Order: Census Cards, Housewifery, and the State's Data Compilation," in Bittel et al., *Working with Paper,* 110.
74 Library Bureau, *Filing as a Profession for Women,* 9.
75 Library Bureau, 7.
76 For a list of the Boston School of Filing Alumnae Association speakers for 1921, see "Boston Filing Association," *LB File,* December 1921, 7, Herkimer County Historical Society.
77 Catherine McNulty, "Warren Filing Association," *Filing,* January 1921, 718.
78 "Filing Association Reports," *The File,* April 1932, 7; "Association Happenings," *Filing and Office Management,* August 1921, 74.
79 Ethel G. Armstrong, "New York Association," *Filing and Office Management,* April 1922, 116.
80 "Filing Association Reports," April 1932, 7; "Filing Association Reports," *The File,* June 1933, 3, Herkimer County Historical Society.
81 For examples, see "Filing Association Reports" in the following issues of *The File* (each on page 3 of its issue; all issues held at the Herkimer County Historical Society): November 1932, February 1933, April 1933, June 1933, and December 1933.
82 Library Bureau, *Filing as a Profession for Women,* 48.
83 Williams, "Documentation and Special Libraries Movements," 777.
84 Quoted in Alistair Black, "Hidden Worlds of the Early Knowledge Economy: Libraries in British Companies before the Middle of the 20th Century," *Journal of Information Science* 30, no. 5 (2004): 432.
85 Williams, "Documentation and Special Libraries Movements," 775; Ethel Vernon Patterson, "Must Work Hand-in-Hand," *Filing,* March 1920, 605.
86 Dorsey W. Hyde Jr., ed., "The Special Library and the Filing Department: A Symposium," *Filing,* March 1920, 602.
87 Williams, "Documentation and Special Libraries Movements," 779.
88 Library Bureau, *Filing as a Profession for Women,* 46.
89 "The Business Library," *Filing,* October 1918, 129.
90 "The Business Library," 129.
91 Strom, *Beyond the Typewriter,* 321.
92 Kwolek-Folland, *Engendering Business,* 11.

Chapter 7. Domestic Storage

1 Montrose J. Moses, "Concerning Home Pockets—A Plea for System," *House and Garden,* April 1930, 76.
2 Paul Jerman, "Keepers Finders," *American Home,* April 1940, 42, 107.

3. Jerman, 107.
4. Lisa Gitelman, *Always Already New: Media, History, and the Data of Culture* (Cambridge, Mass.: MIT Press, 2006), 89–93; Robertson, *Passport in America*.
5. Jerman, "Keepers Finders," 42.
6. Craig Robertson, "Paper, Information, and Identity in 1920s America," *Information & Culture* 50, no. 3 (2015): 394.
7. Robertson, *Passport in America*, 215–44.
8. Jerman, "Keepers Finders," 42.
9. Jerman, 107.
10. Colleen McDannell, *Material Christianity: Religion and Popular Culture in America* (New Haven, Conn.: Yale University Press, 1995), 67.
11. McDannell, 98, 72–73.
12. McDannell, 96.
13. Deidre Lynch, "Paper Slips: Album, Archiving, Accident," *Studies in Romanticism* 57, no. 1 (Spring 2018): 90.
14. McDannell, *Material Christianity*, 89.
15. McDannell, 90–91.
16. Robertson, *Passport in America*, 106.
17. McDannell, *Material Christianity*, 101–2; Katherine C. Grier, *Culture and Comfort: Parlor Making and Middle-Class Identity, 1850–1930* (Washington, D.C.: Smithsonian Books, 2010).
18. Martha Langford, *Suspended Conversations: The Afterlife of Memory in Photographic Albums* (Montreal: McGill University Press, 2008), 124.
19. Langford, 63.
20. However, Spigel dates this move to the middle of the twentieth century, with George Nelson's Storagewall being the key example. Lynn Spigel, "Object Lessons for the Media Home: From Storagewall to Invisible Design," *Public Culture* 24, no. 3 (2012): 571.
21. George Nelson and Henry Wright, *Tomorrow's House* (New York: Simon & Schuster, 1945), 137.
22. Mary Anne Beecher, "A Place for Everything: The Influence of Storage Innovations on Modern American Domesticity (1900–1955)" (PhD diss., University of Iowa, 2003).
23. Gregg, *Counterproductive*, 22–24, 27–28.
24. Gregg, 26–28.
25. Gregg, 33.
26. Gregg, 49.
27. Gregg, 4.
28. Beecher, "Place for Everything," 49.
29. Beecher, 200; Henry Urbach, "Closets, Clothes, Disclosure," *Assemblage*, no. 30 (August 1996): 64.
30. Beecher, "Place for Everything," 189–205.
31. Urbach, "Closets, Clothes, Disclosure," 64, 66.
32. Beecher, "Place for Everything," 196.
33. On retail display, see Beecher "Place for Everything," 160–76.

34 Louis H. Gibson, "Serviceable Closets," *Ladies' Home Journal,* April 1890, 7.
35 "The Inside of the House: Linen Closets and Clothes Closets," *House Beautiful,* September 1915, 121.
36 "Inside of the House"; Anna B. Hill, "A Closet for Every Need," *American Builder,* February 1919, 48–49.
37 Ethel Bartholomew, "Woman's Idea of Convenient Closets," *House Beautiful,* March 1908, 31.
38 Clive Edwards, "*Multum in Parvo*: 'A Place for Everything and Everything in Its Place'; Modernism, Space-Saving Bedroom Furniture and the Compactom Wardrobe," *Journal of Design History* 27, no. 1 (2013): 26–27.
39 Eleanor Raymond, "A Model Kitchen: Designed for the Woman Who Does Her Own Work," *House Beautiful,* June 1932; "This Kitchen Has Five Departments," *House Beautiful,* January 1949, 66–71; Mabel J. Stegner and Josephine Wylie, "Kitchens in Logical Order," *Better Homes and Gardens,* October 1934, 18–19, 100; Grace Pennock, "Better Places to Put Things," *Ladies' Home Journal,* March 1935, 34–35.
40 Elva D. Hoover, "The Useful Kitchen Cabinet: That Consolidates Storage Facilities and Work Centers," *House Beautiful,* July 1924, 552, 598. See also Ella Cushman, "The Development of a Successful Kitchen," *Cornell Bulletin for Homemakers,* June 1936, 7.
41 Mary Anne Beecher, "Promoting the 'Unit Idea': Manufactured Kitchen Cabinets," *APT Bulletin: The Journal of Preservation Technologies* 32, nos. 2–3 (2001): 31.
42 "A Kitchen with Everything in Reach: A Man's Idea of a Conveniently Arranged Kitchen," *Ladies' Home Journal,* May 1914, 91.
43 Clara E. Jonas, *Kitchen Storage Space* (Cornell Extension Bulletin no. 398) (Ithaca, N.Y.: Cornell University Press, 1938), 19–20.
44 Jeffrey Kastner and Deanna Day, "Bringing the Drugstore Home: An Interview with Deanna Day," *Cabinet* 60 (Winter 2015–16): 95–101.
45 Nancy R. Hiller, *The Hoosier Cabinet in Kitchen History* (Bloomington: Indiana University Press, 2009).
46 Beecher, "Place for Everything," 79.
47 Beecher, 31.
48 Beecher, "Promoting the 'Unit Idea,'" 30–31.
49 Beecher, "Place for Everything," 114.
50 Beecher, 125, 126; Hiller, *Hoosier Cabinet in Kitchen History,* 59.
51 Jonas, *Kitchen Storage Space,* 5.
52 Sylvia Comfort Starr, "Office for the Housewife," *House Beautiful,* September 1933, 86.
53 Dianne Harris, *Little White Houses: How the Postwar Home Constructed Race in America* (Minneapolis: University of Minnesota Press, 2013), 199.
54 Harris, 203.
55 Harris, 200.
56 Christine Frederick, *Household Engineering: Scientific Management in the Home* (Chicago: American School of Home Economics, 1920), 305.

57 Christine Frederick, *The New Housekeeping: Efficiency Studies in Home Management* (New York: Doubleday, 1914), 128.
58 Frederick, 126.
59 Frederick, *Household Engineering*, 298.
60 Frederick, *New Housekeeping*, 133.
61 Frederick, 140.
62 Frederick, 144.
63 Frederick, 144; Gregg, *Counterproductive*, 33.
64 Frederick, *New Housekeeping*, 127–28.
65 Moses, "Concerning Home Pockets," 76.
66 For a history of pockets in women's clothing prior to 1900, see Barbara Burman and Ariane Fennetaux, *The Pocket: A Hidden History of Women's Lives, 1660–1900* (New Haven, Conn.: Yale University Press, 2019).
67 Anne Hollander, *Sex and Suits: The Evolution of Modern Dress* (New York: Alfred A. Knopf, 1994), 8.
68 Moses, "Concerning Home Pockets," 76.
69 Moses, 150.
70 Moses, 76.
71 Moses, 76.
72 Moses, 150.
73 Moses, 148.
74 Moses, 150.
75 Spigel, "Object Lessons for the Media Home," 546. Shannon Mattern includes the Storagewall in her history of intellectual furnishings; see Mattern, "Intellectual Furnishing."
76 It was not a complete loss for Moses's critique. As Spigel perceptively notes, at the same time George Nelson was celebrating the information logic of storage, Gaston Bachelard's *The Poetics of Space* was published in France. In this oft-cited book Bachelard argues that home remains a space that "shelters daydreaming [and] . . . protects the dreamer"; cabinets and drawers are places to store secrets and daydreams. Spigel, "Object Lessons for the Media Home," 569.
77 Spigel, 559.
78 Frederick, *New Housekeeping*, 55.

Afterword

1 By the early twentieth century, the term *file cabinet* had largely replaced *filing cabinet*, especially in American usage. In this Afterword I use *file cabinet* when discussing cabinets from the mid-twentieth century onward; I revert to *filing cabinet* when referring to cabinets from the first half of the twentieth century.
2 Susan Green, "Miles of Files: A 'Connected' Art Exhibit Puts Paperwork in Perspective," *Seven Days,* September 25, 2002, https://www.sevendaysvt.com; Pamela Polston, "WTF: What's with the File Cabinet Tower on Burlington's Flynn Avenue?," *Seven Days,* December 16, 2015, https://www

.sevendaysvt.com. Green's article describes the sculpture being put in place in 2001, but all subsequent articles and interviews with Alvarez date the work to 2002.
3. Emily Corwin, "Will Burlington's Champlain Parkway Ever Get Built?," November 15, 2019, Vermont Public Radio, https://www.vpr.org; Jordan Adams, "Very Cool Organizes a File-Cabinet 'Worship' Service," *Seven Days*, April 24, 2019, https://www.sevendaysvt.com; Aidan Quigley, "Federal Decision Delays Champlain Parkway Construction," November 26, 2019, VTDigger, https://vtdigger.org.
4. Kafka, *Demon of Writing*, 78–79.
5. Kafka, 117–18.
6. For analyses of this anxiety in relation to the development of identification practices, see Robertson, *Passport in America*; Lauer, *Creditworthy*.
7. David Fahrenthold, "Sinkhole of Bureaucracy," *Washington Post*, March 22, 2014, https://www.washingtonpost.com; Bob Brewin, "VA's Backlog of Paper Claims Could Cause a Building Collapse," Nextgov, August 10, 2012, https://www.nextgov.com.
8. Nick Levine, "The Nature of the Glut: Information Overload in Postwar America," *History of the Human Sciences* 30, no. 1 (2017): 34–42.
9. Levine, 45.
10. Mills, *White Collar*, 189–212.
11. Richard Mander, Gitta Salomon, and Yin Yin Wong, "A 'Pile' Metaphor for Supporting Casual Organization of Information," in *CHI '92: Proceedings of the SIGCHI Conference on Human Factors in Computing Systems*, ed. Penny Bauersfeld, John Bennett, and Gene Lynch (New York: Association for Computing Machinery, 1992), 627. See also the foundational article in desktop organization research, Thomas Malone, "How Do People Organize Their Desks? Implications for the Design of Office Information Systems," *ACM Transactions on Office Systems* 1, no. 1 (1983): 99–112.
12. Thomas Haigh, "How Data Got Its Base: Information Storage Software in the 1950s and 1960s," *IEEE Annals of the History of Computing* 31, no. 4 (2009): 21; Steven Johnson, *Interface Culture: How New Technology Transforms the Way We Create and Communicate* (New York: Harper Edge, 1997), 42–50.
13. Victor Kaptelinin and Mary Czerwinski, "Introduction: The Desktop Metaphor and New Uses of Technology," in *Beyond the Desktop Metaphor: Designing Integrated Digital Work Environments*, ed. Victor Kaptelinin and Mary Czerwinski (Cambridge, Mass.: MIT Press, 2007), 1.
14. Edward W. Briss, "The Integrated Software and User Interface of Apple's Lisa," in *AFIPS '84: Proceedings of the July 9–12, 1984, National Computer Conference*, ed. Dennis J. Frailey (New York: Association for Computing Machinery, 1984), 319–28; Nathan Lineback, "Magic Desk I for Commodore 64," Nathan's Toasty Technology Page, accessed September 17, 2020, http://toastytech.com/guis/magdesk.html; "Magic Desk I—Commodore 64 (NTSC) [Mess] [Shortplay]," YouTube, posted January 4, 2016, https://www.youtube.com/watch?v=nTnJs40Loo8.

15 Kaptelinin and Czerwinski, "Introduction," 2.
16 Mander et al., "'Pile' Metaphor," 627.
17 More advanced computer users can use file aliases to store documents in multiple places.
18 Ted Nelson, "The Unfinished Revolution and Xanadu," *ACM Computer Surveys* 31, no. 4es (December 1999): 5.
19 Jaron Lanier makes a similar argument: "Our conception of files may be more persistent than our ideas about nature. I can imagine that someday physicists might tell us that it is time to stop believing in photons, because they have discovered a better way to think about light—but the file will likely live on. The file is a set of philosophical ideas made into eternal flesh." Jaron Lanier, *You Are Not a Gadget: A Manifesto* (New York: Alfred A. Knopf, 2010), 13.
20 Eric Freeman and David Gelernter, "Beyond Lifestreams: The Inevitable Demise of the Desktop Metaphor," in Kaptelinin and Czerwinski, *Beyond the Desktop Metaphor*, 43.
21 King Features Syndicate, "Barney, the File Clerk, Gets Fired," 42, 43.

Index

Page numbers in italics refer to illustrations.

access, 43, 49, 58, 64, 96–98, 99, 101, 103, 107, 108, 110, 145, 173, *205*; efficiencies of, 24, 62; facilitating, 36–37, 96, 114; focus on, 148; integrity and, 80; physical, 233

accumulation, 23, 35, 41, 42, 63, 126, 188, 274n64

accuracy, 69, 166, 168, 195

adding machines, 210

addressing machines, 189

affordances, 3, 4, 257, 266n6, 268n23

Agar, Jon, 139, 140

Alvarez, Bren, 251, 253, 299n2

American Documentation Institute (ADI), 219

American Library Association (ALA), 219

American Management Association, 142

Apple, Inc.: desktop metaphor and, 255–56; icons and, 22; information management and, 255

Art Metal, 9, 82, 84, 96, 112, 161; advertisement by, *5, 47, 83, 88–89, 90, 141, 154, 192*; plan files by, 154; terrasbestos and, 92

asbestos, 92, 93

attachment, 64, 68, 70, 80, 101, 124. *See also* binders

Bachelard, Gaston, 298n76

ball bearings, 6, 98, 143

Beecher, Mary Anne, 239

Beniger, James, 269n32, 283n3

Bibles, 228–31, 232; family memories and, 228–29; as storage containers, 230

binders, 66, 69, 123, 125. *See also* ledgers: loose-leaf

binding, 18, 64, 67–68, 73

Blair, Ann: on information, 268–69n27

blueprints, 153, *155*, 237; Cello-Clip Map and Plan File, 154; Mammoth Vertical File, 154, *155*; storing, 154, *154*

bookkeeping, 49, 132, 180, 293n46; machines, 189, 210; status of, 294n67

books, 49, 77, 86, 125, 152, 189, 209; account, 67, 69, 256; bound, x, 67, 72, 275n8; copy, ix, 65; filing, 114; integrity of, 64, 70; ledgers and, 68–69, 152; letter, 25; model for reading file, 123; ontological

301

conceptions of, 65; press, ix; repurposing, 67; signature, 152; as storage technology, 18, 65, 69; structure of, 67. *See also* page
Boston School of Filing, 217, 295n76
boxes, 44, 75, 170, 234; cardboard, 71, 72; partitioned, 105
Braverman, Harry, 185
"Built Like a Skyscraper" advertising campaign, 11, 12, 36, 37, 82, 86, 96, 186, 189. *See also* Shaw-Walker
bureaucracy, 1, 201, 225; government, 254; as-paperwork, 254
Bush, Vannevar, 288n19
business: expansion/growth of, 36, 53; vitals of, 65

cabinet logic, 25, 26, 77, 104, 105, 108, 110, 115, 132, 134, 164, *184*, 223, 228, 232, 234–35, 248, 279n5
cabinets, 6, 75, 77; card, 132, 157; counter-filing, 59; drawers and, 98; flat file, 154; index card, *146, 147*, 290n58; large, *79, 81*; memory, 247; partitioned, 105; as physical structure, 23; stand-alone, 36; supply, 142; wooden, 73, 77, *78*. *See also* card catalog cabinets; file cabinets; filing cabinets; kitchen cabinets
calculators, 210, 256
capitalism, 126; corporate, 4, 23, 25, 33, 35, 36, 37, 134, 157, 168, 197; filing cabinet and, xiii, 3, 257; gender and, 23, 248; government and, 3; industrial, 188; information and, 3, 258; labor and, 176; logics of, 36, 59, 62; managerial, 135; modernity and, 36; skyscrapers and, 41; vertical bias of, 33, 35, 62. *See also* efficiency; granular certainty
carbon copies, 70, 211
card catalog cabinets, 79, 111, 123, 157, 290n58

card indexes, 106, 242, 243, 279n8
cards, 117, 145, 147, 148, 158; cross-reference, 151; filing, 143; ledger, 152; manila, 151; placement of, 132; standardized, 185. *See also* guide cards; index cards
Carnegie, Andrew, ix, 35
Carruthers, Mary, 170, 172
case/book box *(scrinium)*, 170
cells, 43, 44, 75, 92, 170. *See also* granular certainty
centralization, 28, 42, 50, 142, 145, 226, 244
Cevolini, Alberto, 174, 289n32
charge cards, 166, *167*
check protectors, 210
chronological order, ix, 67, 124, 289n33
circulation, 35, 45, 105, 110, 120. *See also* flow
class, 26, 27, 195; personality and, 203–4, 206–7. *See also* middle class
classification, x, 20, 110, 132, 136, 142, 170, 214; definitions of, 280n8; formal, 255; teaching, 211; as technology, 168–69; as temporal problem, 18, 105, 108
classification systems, 101, 104, 121, 132, 150, 211, 217. *See also* chronological order; filing systems; indexing
clerical work, 186, 199, 239, 255; apprenticeship model of, 50; brain work in, 209; emergence of, 185; ideal mode of, 195; pay for, 198; as profession, 220; thought in, 209; women and, 4, 25–26, 197, 292n10. *See also* filing; office work
clerical workers, 292n10; age of, 199; female, 12, 206; filing cabinets and, 167. *See also* file clerks; office workers
clippings, 152; storage of, 150–51
closets, 26, 223, 224, 232–35
clothing, 79, 199, 202, 232, 233, 234, 244; work, 204, 206, 207

Commission on Economy and Efficiency (Taft Commission), 147, 148
Commodore International, 256
compartmentalization, 75, 233, 237, 257
compression, 110, 111, 114, *154*. *See also* decompression; Wobble Block
compressors, 108–14, *112*; enameled steel plate, 113; locked/released, 109. *See also* Wobble Block
computers, xii, 22, 254–55, 256, 257
congestion, 53, 104, 126, 127; culture of, 41; verticality and, 42
containers, 8, 33, 65, 101, 143, 237, 239, 257, 270n45; storage, 23, 82, 230, 245
continuity, 69, 70, 103, 177, 180; storage principle, 23, 37, 63
control. *See* efficiency
correspondence, 54, 117, 123, 124; collection of, 24; storing, ix–x, 152
Corwin, Sharon, 290n55
Couldry, Nick, 4
counters, 59, *60, 61*
credit reports, 3; index for, 150
cross-referencing, 151, 174, 272n10, 289n33, 289n34, 289n37
Csiszar, Alex, 20
cupboards, 72, 228, 235, 237

data, 170, 185, 264; big, 4, 6; digital, 3, 22, 27; files and, 20, 22, 23, 27; information and, 16, 132, 145, 147, 242, 269n28; knowledge and, 20, 145, 269n28; materiality of, 27; tabulating/compiling, 295n73
Day, Deanna, 237
Day, Ronald, 64
decompression, 114
Department of State, x, 18, 63, 148, 229; filing system by, ix
desks, 28, 44, 45, 46, 62, 145, 207, 233, 256; desk diseases, analysis of, 188; Desks with Brains, 170; drawers, 72, 226; "Efficiency Desk," 180, *181,* 189, 224; and filing cabinets, 49; "flat top," 48, *48, 50, 171*; kitchen, 241–42; nineteenth-century, 49; storage and, 48, 49, 50, 72, 73, 142, 209, 255. *See also* Wooton Desk Company
desktop metaphor, 3, 22, 255–56, 257
Dewey, Melvil, 8
Dewey decimal system, 148
dexterity, female, 185–86, 188, 195, 216
Direct Name Index, 166
domesticity, 197, 221, 229
domestic managers, 223, 224
drawer fronts, 72, 73, 75, 99, 278n58
drawers, 25, 35, 49, 50, 75, 99, 103–4, 105, 117, 124, 138, 145, 151, 153, 175, 180; cabinets and, 98; in closets, 234, 235; file, 73, 113, 170; horizontal, 77, 79; in kitchens, 235, 237, 239; labels on, 78; locking, 95, 96; long/short, 234; movement of, 79, 90, 96, 97, 98; opening/closing of, *11,* 12–13, 53, 84, 98, 101, 112; partitioning of, 157, 182, 233, 237, 239; stacking of, 33, 36; wooden, 97; wooden files used as, 77, 78, 79, 80; working, *184*
drawer slides, xii, 6, 87, 96–98; diecast rollers and, *98*; importance of, *97,* 98

Edison, Thomas, 84
education, 26, 198; commercial, 207, 208, 209
efficiency, xi, 43, 51, 66, 69, 84, 119, 135, 140, 158–59, 177, 231, 253; celebration of, 156, 164; control and, 23, 37, 128, 134, 136, 138, 139, 168, 169, *178*; discourses of, 23, 105, 176; filing and, *119,* 128, 143, 147, 163, 204; gender and, 11, 23, 25, 56, 193, 203, 224, 259; in home office, 224, 228, 241; ideas of, 117, 120; imagination and, 33; importance of, 17, 18, 64, 141; information and, xiii, 6, 16–18, 20,

22–28, 46, 121, 134, 158, 161, 168, 173, 193, 249, 269n32; machine, 143; material history of, 4, 16, 33, 37, 249; middle-class values and, 204; particular/discrete and, 17, 67, 115, 117, 132, 136, 232, 235, 237, 239; principle of, 31, 41, 70, 103, 108, 120, 242; productivity and, 46, 99, 101, 105, 134, 158, 168, 223, 228, 232, 245, 246, 248; promoting, 156; spatial organization and, 223; speed and, 166, 248; system and, 4, 8; temporal, 110; understanding, 23, 224; verticality and, 33, 41, 53, 56, 103, 108

elevators, 42, 43, 44, 45

Emerson, Harrington, 242

Empire State Building, 37

enclosure, 23, 24, 37, 63, 77, 101; technology of, 68–69

engineering, xii, 153; management and, 135

envelopes, 9, 41, 43, 152, 154, 244; file, 124, 234; manila, 151

equipment. *See* office equipment

Ernst, Wolfgang, 22

error, 98, 166, 201; possibility of, 136

expansion, 53, 70, 101, 114, 116, 127, 151

fasteners, 124, 125

Fenske, Gail, 37

file cabinets, 12, 225, 228, 251, 253, 254, 257, 259; distinction from filing cabinet, 14, 226, 298n1; icon, 255–56

file clerks, xi, 22, 45, 51, 115, 120, 126, 142–43, 185, 187, 259; age, 195, 199, 201–3; allocation of space and, 58; class identity and, 203–4; clothing and, 207; described, 195; difference from file executive, 215, 218; height of, 53–54; men as, 189, 190, *191*; by numbers, 198–99; role of, 242; test for position of, 216; women as, 55, 182, 189, 193, *196*, 198–203, 215–16, 220–21. *See also* clerical workers

file executives, 142, 215, 216, 218; described, 220–21

file girls, 199, 202

files, 45, 77; accordion, 127; administrative, 120; alphabetical, 151; board, 71, 73, *74*; box, 64, 71, 72, *72*, 73; card, 147, 242; case, 3, 121; computer, 3, 22, 27, 255, 270n43; document, 71, 72, 75, *76*, 148; flat, 71, 72, 73, *74*, 75, 78, 81, 148, 153, 154; history of, 120; icon, 255–56; as information formats, 127; legal, 124; slumping, 114; suspension, 80; as technology of gathering, 120; transfer of, 217; vertical, 7–8, 70; vertical (library), 151, 152, 153, 157. *See also* gathering

filing: automatic, 168, 175–76, 182, 183, 190, 201; commercial education and, 209; informal training in, 186; as information labor, 26, 163, 176, 220; profession of, 215–21; teaching, 207–11, 214; women and, 26, 178, 221. *See also* dexterity; file clerks; hands; vertical filing

Filing (magazine), 217, 219

filing associations, 217, 218, 220

filing cabinets: bases, 51, *52*, 56; as counter/divider, 59, *61*; disorganized, 144–45; distinction from file cabinet, 14, 226, 298n1; emergence of, 27–28, 36, 70, 92, 246; fire and, 91–92, 93, *94*, 95; five-drawer, 53, 55; four-drawer, *2*, 51, 53, *54*, 55, 91; height of, 32, 33, 51, 53, 55, 56; in home, 225–26, 228–31; integrity of, 63, 82, 84–87, 90–92, 94–99, 101, 173; invention of, 8–9, 70, 267n15; locks, 95–96; as a machine, 8, 26, 163–68, 173, 175, 180, 183, 185, 190, 201, 220, 226; manufacture of, 7, 12, 83, 86–91; modular, 51, 53, 56, 272n10;

nineteenth-century, 70, 75, 77, 79, 80, 82; operating, *5, 13, 15,* 51, 55, *141, 155, 174, 183, 187,* 190, *191, 192, 196, 200, 205, 221, 258*; patents, 6, 9, 109, 267n17; placement, in office, 51, 53, 56, 58–59; price, 91; steel, 82, 84, 85, 86–87, *90,* 91, 92, 101, 278n62; three-drawer, *60*; vanishing point of utility of, 51; wooden, 77, 84, 87, 90, 91, 97, 277n56. *See also* file cabinets; horizontal filing cabinets; vertical filing cabinets

filing department, 50, 142, 143, 195, 214, 218, 220

filing equipment. *See* compressors; folders; guide cards; tabs

filing equipment companies, 9–10, 114, 124, 127, 150, 217; factories, 9; patents and, 9. *See also* Art Metal; Flexifile; Fred Macey; General Fireproofing; Globe Company; Globe-Wernicke; Library Bureau; Remington Rand; Safe-Cabinet Company; Schlicht and Field; Steelcase; Vawter-Baker; Wabash Cabinet; Yawman and Erbe

filing systems, 3, 23, 59, 80, 127, 140, 142, 145, 157, 167, 218, 259; alphabetical, 99, 115, 143, 147, 150, 151, 152, 164, 166, 211; alphanumeric, 99, 164, *165*; decimal, x, 148, 158; geographical, 143, 211; numerical, ix, 99, 164, 166; subject, ix, x, 7, 148, 150, 152, 156, 211, 214, 215, 219, 220. *See also* classification; indexing

Fine, Lisa, 199, 201

Flexifile: Desk with Brains, 170; filing devices from, *171*

flow, 24, 33, 36, 37, 43, 45–46, 59, 62; of information, 46, 50, 139, 140, 163, 254; of paper, 48, 50, 51, 64, 139; of people, 41, 42, 273n35; storage and, 48; of work, 46, 103, 139. *See also* movement; workflow

folders, xii, 35, 123–28, 138, 152; accordion/expansion, 127, 151; attaching paper to, 124; as enclosure, 20, 120, 121, 123, 125, 126; envelope, 124; as file, 120–21; hanging, 109, 111; horizontal plane of, 117; icon, 3, 22, 27, 256, 257; information and, 3, 14, 20, 25, 27, 121, 128, 179; manila, 14, 16, 18, 22, 25, 27, 70, 105, 108, 117, *118,* 120, 121, *122,* 124, 125, 126, 128, 164, 179, 185, 248, 255, 257, 259; misuse of, 126, 127, 179; as partition, 105, 108, 115, 120, 127, 164, 185; removing, 25, 114, 115, 123, 179; sizes, 123; sulfate, 126, 282n62; thickness, 126; in vertical filing, 14, *21,* 77, 121. *See also* gathering; guide cards; tabs

Frederick, Christine, 242, 243, 247, 248. *See also* scientific management

Fred Macey, 9; advertisement by, *21*

furniture, 224, 228, 240, 241; steel, 84; storage and, 231

Galloway, Lee, 195, 197, 198

Gardey, Delphine, 140, 180

gathering, 120–21, 123–28, 175, 220, 244. *See also* files

Geertz, Clifford, xii, 265n5

gender, xiii, 13; capitalism and, 248; class, race, sexuality, age and, 23, 26, 195, 220, 224, 291n3; efficiency and, 11, 23, 25, 56, 193, 203, 224, 259; information and, 14, 25, 27, 177, 249, 259; information labor and, 26, 176, 177, 180–90; labor and, 3, 4, 11, 27, 176, 201, 259, 294n67; manual and mental work and, 12, 26, 58, 112, 180, 209; office space and, 199, 201; office work and, 11–12, 14, 112, 163, 189, 190, 192, 193, 195, 254, 258, 291n3; typewriter and, 177

General Fireproofing, 9, 84; catalog,

57, 109, 118, 183; GF Allsteel Files, 91; Super-Filer, 183
General Motors, 10
Gilbert, Cass, 40
Gilbreth, Frank, 290n55
Gilbreth, Lillian, 290n55
Gitelman, Lisa, 163, 169, 177, 269n34
Globe Company, 77; catalog, 76, 78; Globe Art, filing cabinet from, 2
Globe-Wernicke, 114, 117, 143–44, 154, 190; catalog, 99, 174, 187, 191; Safe-Guard, 166; Tri-Guard files, 174, 175; vertical filing cabinet invention and, 9
Google, Mr., 258, 259
granular certainty, 17, 18, 121, 136, 137, 157, 163, 248; cabinet logic and, 128, 134, 148; closets and, 234; defined, 3, 132; information labor and, 177; kitchen cabinets and, 239; memory and, 173, 231; storage and, 151, 170, 224, 246, 247; system and, 26, 134, 158, 173
Gregg, Melissa, 231, 232
guide cards, 7, 25, 105, 108, 112, 115, 116, 117, 118–19, 132, 164, 173, 174, 185, 186, 211

hands: cabinets and, 105; disembodied, 182, 184, 185, 187, 192; grasping, 105, 179, 185, 193; handling paper, 8, 111, 115, 166, 179, 193, 202; "hired hands," 182; machines and, 177, 185; men's, 184, 188, 189; women's, 186, 220. See also dexterity
Harris, Dianne, 241
hierarchy, 39, 58, 163, 176, 197, 224; clerical, 204, 220; management and, 35; office, 180, 189, 254
Higgins, Hannah, 43, 44
high heels, 55, 56, 205, 274n76
Hollander, Anne, 244
home: office and, 223, 241–43, 243–49; organization in, 224; storage and, 223–49

Hoosier Company: advertising budget for, 241; kitchen cabinets, 237, 238, 239, 241, 247, 248
horizontal filing cabinets, 27, 153, 157, 233; image of, 100; influence on kitchen cabinets, 237; and vertical filing cabinets compared, 99
horizontality, 35, 36–37, 39, 40, 41, 45, 62, 117

index cards, 22, 56, 62, 64, 99, 111, 133, 190, 275n8, 280n8; filing of, 46, 132, 143, 145; information on, 131, 132, 147, 242; memory cabinet of, 247; retrieval of, 184, 185; storing (Visible Index), 149; using, 131
indexes, 70, 106, 148, 166, 279n8; alphabetical, 105, 137, 150, 174, 214; automatic, 105, 150, 164, 165, 166; definition of, 285n47. See also visible indexes
indexing, 150, 185, 218, 285n47; classification and, x, 105, 211, 279n8; definitions of, 280n8; projections (tabs), 115; science of, 215, 217; search and, 148; systems, x, 104, 114, 121, 132, 140, 174. See also classification; cross-referencing; filing systems
industrialization, 43, 59, 62, 85, 186, 188
inefficiency, 53, 55, 56, 59, 147, 150, 153, 169, 247, 175, 253
information, 4, 6, 218–19, 229, 230; ascendancy of, 16, 27, 151, 249; capitalism and, 258, 259; conception of, 14, 16, 18, 20, 28, 128, 132, 134, 151, 157, 169, 177, 220, 249, 269n28; data and, 16, 132, 145, 147, 242, 269n28; efficiency and, 16–18, 20, 22–28, 46, 128, 134, 173, 177, 249, 269n32; experiencing, 25, 214, 224, 228, 247; at fingertips, xiii, 25, 115, 179, 185, 193, 225, 226, 228, 259; flow of, 42, 46, 48, 64, 105, 134, 139, 140, 168, 188,

254, 257; granular certainty and, 17, 18, 121, 128, 132, 134, 136, 137, 148, 157, 158, 163; handling, 134, 185, 192–93; knowledge and, 27, 179, 180, 266n7; logic, 151, 247; materiality of, 27, 127; paper and understanding of, xiii, 14, 16, 18, 20–22, 23, 27, 64, 121, 132, 136, 138, 158, 179, 219, 224, 242; productivity and, 17, 18, 46, 134, 136, 139, 218; retrieval of, 3, 14, 50, 67, 115, 149, 151, 159, 190, 242, 243, 247, 249; storing, 3, 105, 128, 147, 226, 255, 257, 288n19; vertical history of, xiii, 27–28

information labor, 25, 26, 163, 176–77, 179–80, 182, 185–86, 188–90, 192–93, 195, 215, 220

information overload, 254, 259

information theory, 17, 269n29

integrity, 23, 25, 78, 125, 128; access and, 64, 80, 96, 99; of book, 64, 65, 67, 68, 70; completeness and, 70, 77, 173; definition of, 63, 67, 95; family Bible, 229; filing cabinet, 24, 63–64, 70, 82, 84–87, 90–92, 94–99, 101, 103, 140, 173, 226, 253; of loose-leaf binder, 65–70; material wholeness and, 68; moral, 68; principle of storage and, 23, 63, 64, 68, 120, 128; protection and, 78–79, 86, 91, 92, 94, 96, 99; storage and, 65, 68, 77, 80, 123, 124; structural, 64, 67, 77, 103

Jerman, Paul, 226–27, 230, 241, 247; filing cabinets/home and, 225; information and, 228

Kafka, Ben, 197, 201, 253

kitchen cabinets, 26, 224, 233, 235, 237, 241; Kitchen Maid, advertisement for, *236*; Napanee Dutch Kitchenet, 239, *240*; Seller's, advertisement for, 239. *See also* Hoosier Company

kitchen design, 235, 236, 240

knowledge, 159, 174, 176, 214, 231, 259; data and, 20, 145, 269n78; information and, 4, 16, 27, 179, 180, 266n7; instrumental (information), 4, 6, 17, 145, 228, 249; paper and, 20

Krajewski, Markus, 67, 174

Krause, Louise, 274n64

labor: capitalism and, 35, 176; class and, 26, 215, 220, 242, 291n3, 295n73; efficiency and, 128, 177, 246, 248, 290n55; expense of, 53, 55, 203; gendered, 177, 180, 186, 201, 224, 259, 294n67; machine, 136, 175, 176, 185, 201; manual, 185, 189, 193, 209, 239; mental, 209; technology and, 182. *See also* information labor

Lange, Alexandra, 271n4, 273n49, 273n51

Langford, Martha, 230

Lanier, Jaron, 300n19

Lapartito, Kenneth, 291n64

Larkin Building, 46

Le Corbusier, 271n4

ledgers: books and, 68–69, 152; bound, 25; cards, 152; loose-leaf, 64, 65–70, *66,* 80, 101, 275n8

Leffingwell, William Henry, 31, 33, 108, 121, 158, 159; death of, 138; efficiency principle and, 41; filing cabinet drawers and, 103–4; retrieval and, 104; simplification and, 140; verticality and, 56, 104. *See also* scientific office management

letters, 138, *212, 213,* 225; of the alphabet, 72, 142–43, 214; filing, 51, 73, 104, 108, 170, 190, 211, 213, 214; love, 245; sorting, 143; storage of, x, 2, 18, 245, 247; typewriter keys and, 177

Levine, Nick, 254

librarians: company/business, 20, 218, 219, 220; special, 218–19

libraries: company/business, 50, 218, 219; public, 151, 219; vertical file, 151–52
Library Bureau, 84, 99, 108–9, 123, 150, 166, 211, 267n17; catalog, *2, 97, 106, 122, 133, 167, 221*; factory, *10*; *Filing as a Profession for Women,* 215–16; history of, 267n15; pamphlet, *60, 61, 107*; teaching equipment, 210, *210,* 211, *212, 213*; vertical filing cabinet invention and, 8–9
Litterer, Joseph, 135
Liu, Alan, 293n33
locks, 6, 95, 96, 111, 112
Luhmann, Niklas, 174

machines, 135–36, 139–40, 163, 224, 248. *See also* filing cabinets; office machines; typewriters
Magic Desk I, 256
management, 26, 31, 35, 105, 139–40, 168, 204; engineering and, 135; information, 219, 255, 259; literature on, 53, 156, 158, 161; profession of, 17, 134, 135
Martin, Michèle, 294n62
Martin, Reinhold, 271n6
Marx, Karl, 176
materiality, xi, xii, 15–16, 27, 127
Mattern, Shannon, 14, 279n5, 298n75
McCord, James, 157, 179
McDannell, Colleen, 229, 230
McGaw, Judith, 186, 204
Meister, Anna-Maria, 138
memo, 138
memory, 20, 22, 189, 215, 231; arts of, 171, 172, 288n29; automatic, 26, 163, 168–70, 172–76, 235; cross-referencing and, 173–74; delegating, 182, 243; family, 243; filing cabinets and, 161, 164, 168, 172, 173; logistical, 231, 232, 241; office, 139, 168; personal, 139, 230; photographic, 172, 173; sentimental, 231; storage and, 170, 172, 228, 230, 231, 245; as storehouse, 169–70. *See also* recall
middle class, 197, 203, 204, 208, 215, 220
Mills, C. Wright, 272n10; "the enormous file" and, 35, 254
Misa, Thomas, 85
misfiling, 64, 127, 175, 201, 202, 203, 204, 207, 214. *See also* paper: loose
modernity, 34, 151, 226; capitalism and, 36; masculinity and, 11; rational, 231; skyscrapers and, 36, 39, 41
modularity, 43, 51, 99
money pouch *(sacculus),* 170
movement, 36, 43, 45, 46, 48, 101, 105, 139, 140, 224, 235. *See also* flow
Müller-Wille, Staffan, 20
Multisort, 143, *144*
Mumford, Lewis, 23

National Board of Fire Underwriters, 95
National Letter File, *72*
negatives, filing, 285–86n53
Nelson, George, 231, 298n76; active storage and, 246, 247; Storagewall and, 247, 296n20
Nelson, Ted, 257
New York Filing Association, 144, 217
New York Municipal Library, 151
New York School of Filing, 179
Noble, David, 135
Nunberg, Geoffrey, 269n34
Nye, David, 39

office buildings, 36, 85, 273n49
office equipment, 11, 12, 152, 170, 223; difference from office furniture, 9, 62, 84, 164, 239; manufacture of, *88–89*; steel, 84–85, *88–89,* 90, 94
office furniture, 123, 156; difference from office equipment, 9, 62, 84, 164, 239; impact of, 273n51;

placement of, 273n51; work flow and, 48
office machines, 9, 25, 46, 139, 175, 188–89, 204, 210, 291n73; bookkeeping, 180; punch card, 132; women's dexterity and, 25, 185–86, 210. *See also* typewriters
offices, 22, 43–45, 62; commercial, 156; design of, 47, 49, 59, 273n51; gendered division of, 14, 180, 223; home and, 223, 241, 243–49; influence of, 234; modern, 101, 181; organization of, 47, 255; system and, 26
office work, 50, 62; dressing for, *205*, 206; expansion/specialization of, 203; gendered division of, 163, 189, 195, 198, 206. *See also* clerical work
office workers, 42, 50, 189, 192, 208; women, 197, 199, 202, 207. *See also* clerical workers
Olwell, Victoria, 201
Ong, Walter, 65
organization, 40, 67, 128, 134, 170, 172–73, 233, 247, 268–69n27; desk, 47, 255; hierarchy and, 35, 39; home, 223, 224; labor and, 35, 211, 245, 246, 248, 254; movement, 36; office, 47, 59, 145, 189–90, 195, 254, 255; spatial, 24, 31, 36, 105, 110, 223; storage, 35, 148, 223–24, 248; technology and, 140, 148, 156; vertical, 31, 33, 35, 59
Otis, Elisha, 42
Otlet, Paul, 20, 64, 219
out cards, 166

page, ix, 18, 20, 64–70, 72, 73, 125, 229, 230
paper: buckling/crumpling, 126; fastened, 65, 72; flow of, 36, 46, 50, 51, 64, 110; handling, 63, 104, 105, 136, 179, 192–93, 211; information and, xiii, 14, 16, 18, 20–22, 23, 27, 64, 121, 132, 136, 138, 158, 179, 219, 224, 242; loose, 20, 63–64, 65, 68, 69, 72, 77, 78, 96, 124, *125*, 126, 173, 259; manila, 25, 72, 77, 117, 126; misplaced, 25, 64, 65, 73, 201, 204, 214; piles of, 73, 105, 254–57; protecting, 78, 90–92, 94–97, 125, 126, 156; retrieving, 20, 36, 55, 62, 80, 103, 104, 148, 173; size of, 49, 75, 99, 110, 123, 126, 138, 151, 152; storing, 3, 7, 14, 20, 23, 35, 36, 49, 62, 64, 77, 103, 104, 110, 120, 125, 132, 145, 168, 173, 225, 226, 228; unbound, 68, 70, 104, 121, 153, 173; vertical, 7, 25, 103, 108–14, 173
paperwork, 6, 46, 50, 131, 134, 197, 201, 204, 241, 253, 254
particularization, 4, 17, 115–18, 120, 157, 249. *See also* granular certainty
partitions, 25, 49, 59, 75, 77, 101, 105, 157, 164, 223, 232, 237, 248. *See also* cabinet logic
pasteboard, 115, 117, 120
permanence: provisional, 69, 82, 101; as storage principle, 64, 69, 86, 91
personality, 150, 176, 208, 209, 217, 232; gender and, 202, 203–4, 206–7, 215
Peters, John Durham, 101, 177, 268n24, 269n32
photographs, 3, 172; storage of, 150–51. *See also* negatives
pigeonholes, 49, 50, *71,* 72, 77, 105, 153, 170, 172, 234
pile research, 254–55, 257
piling, 24, 55, 110, 255, 259; difference from stacking, 37
Piper, Andrew, 65
planning, 18, 134, 158–59, 168, 216, 218. *See also* productivity
platforms, 56, 274n76
Plato, 170
Plotnick, Rachel, 185
pockets: clothing, 298n66; house,

243–45, 246, 247; paper storage, 125, 143, 151, 154
productivity, 43, 46, 62; efficiency and, 17, 24, 99, 223, 228, 245, 246, 248; filing and, 180, 218, 246; filing cabinet and, 8, 99, 134, 159; home and, 243; information and, 17, 18, 46, 134, 136, 139, 218; self-worth and, 242, 243, 248
protection, 63, 78, 86; filing cabinet, 91–92, 94; folders, 125, 126; locks, 96, 99; as principle of storage, 23, 63, 91
punch cards, 22, 145, 255

race, 26, 27, 195, 197, 220, 224, 291n3
rationality, 17, 204, 211
reading, 58, 117, 123, 156, 210, 211–12; discontinuous, 70; informational mode of, 20; operationalized, 211
recall, 168, 169, 170, 172, 229. *See also* memory
records, 253; business, 69, 139; household, 225, 228, 230, 242, 247; management, 148, 280n8; medical, 145; official, 229; retention, 140
Remington Rand, 10; catalog, *32*; Multisort, 143, *144*; pamphlet, *54, 55, 116, 119*
retrieval, 36, 114, 159, 168, 224, 242, 271n46; cabinet logic and, 104, 110, 115, 151, 157, 184, 223, 235, 247; integrity and, 50, 63, 64, 78, 82, 96, 99, 226; rapid, 18, 24, 80, 81, 110, 115, 148, 218, 247, 248, 249; as storage problem, 18, 24, 35, 49, 111, 173
rollers, 6, 111; die-cast, 98, *98*
rolodex, 256
Root, Elihu, ix, x, 18, 63, 70, 148
Rosenthal, Caitlin, 135
route slips, 166
routine: system and, 69, 204, 209; teaching, 209, 211

Safe-Cabinet Company, 9, 92; catalog, *94, 227*; Furnace Test and, 94–95; storage equipment from, 94
sanitary legs, 51, *52*
Schlicht and Field, catalog, *74, 79*
Scholfield, Ethel, 168, 169, 173, 288n19
scientific housekeeping, 224, 242–43, 245
scientific management, 3, 4, 5, 17, 53, 193; home and, 235, 236, 238; office and, 26, 45, 136–37, 158, 273n55, 284n17; systematic management and, 135. *See also* scientific housekeeping; scientific office management; systematic management; Taylor, Frederick Winslow; Taylorism
scientific office management, 31, 103, 131, 134, 136, 138, 195; impact on office, 284n17; simplification and, 140, 145, 158, 185. *See also* Leffingwell, William Henry; scientific management
Sears, Roebuck, 44
Seibels, Edwin, 9
Shannon Files, 73, 74, 75
Shaw-Walker, 9, 11, 38, 51, 112, 190–91; advertisement by, 12, 36, 37, 82, 96, 186, 189; cabinet/efficiency of, 185; catalog, *11, 13, 34,* 53, *112, 137, 184*; Super-Ideal Index of, *165,* 166; Wobble Block and, 113, *113*. *See also* "Built Like a Skyscraper" advertising campaign
shelves, 14, 49, 56, 72, 73, 75, 123, 156, 228, 246; in closets, 233, 234, 235; in homes, 231; in kitchen cabinets, 235, 237, 239, 245
skyscrapers, 11, 39–43, 44; capitalism and, 41; construction of, 12, 37, 86; division of, 39–40; emergence of, 36; exterior of, 43; filing cabinets and, 4, 24, 33, 34, 62, 82,

86, 92; modernity and, 39; verticality of, 39, 42, 45
space, 18, 22, 44, 75; designed, 234; domestic, 223, 228, 230, 232, 246; drawer, 103, 127; filing, 59, 112, 115, 124; floor, 62, 99; office, 1, 24, 26, 33, 40, 42, 43, 45, 51, 53, 56, 58, 59, 61, 176, 192, 204; organizing, 24, 35, 49, 58, 59; saving, 31, 33, 35, 55, 59, 103, 109, 148; unused, 56. *See also* cabinet logic; compression
Special Library Association, 218–19
speed, 17, 67, 108, 168, 186, 247; efficiency and, 26, 166, 248; filing and, 8, 140, 141, 143, 164, 176, 180; filing cabinets and, 4, 8, 105, 163, 173, 253; work and, 4, 140, 209
Spigel, Lynn, 231, 247, 296n20
spikes, 49, *71,* 72
spindles, 49, 72, 77
standardization, 4, 43, 123, 135, 138, 140, 148, 175, 185, 231; of information, 18, 145, 177, 179
steel, xii, 6, 24, 82, 87, 88–89, 90–91, 113, 251; Bessemer method, 85, 86; development of, 85–86, 277n52; efficiency of, 84, 97; manufacture of, 83, 85, 86; as protection, 91–92; varieties of, 86, 87
Steelcase, 9, 98; advertisement from, *52*
stenographers, 45, 292n14, 293n46
storage, x, 4, 27, 49, 50, 72, 75, 78, 82, 101, 132, 138, 153, 169, 170, 173, 180, 230, 239; active, 246, 247; book and, 18, 64, 65, 69, 70; dead, 105, 108, 158; home, 233–34, 246; kitchen, 235, 241; memory and, 169, 170, 172, 228, 230, 231, 232, 241, 244–45; modern, 140, 228, 246, 248; office desks and, 49; passive, 49, 108; planned, 26, 27, 223, 224, 231–35, 237, 239, 241, 247, 249; principles of, 23, 63, 68, 128, 232; retrieval and, 24, 35, 48, 108, 110, 111; sedentary, 48; values attached to, 23, 27, 63, 65, 108, 128, 134, 224, 228, 232, 246, 248. *See also* cabinet logic
storage space. *See* cabinet logic; storage
Storagewall, 246, 247, 296n20, 298n75
Strom, Sharon, 180
subdivisions, ix, 18, 77, 137, 150, 157, 164, 169, 214
substitution cards, 166
Sullivan, Louis, 40, 44
sway block, 113
System: The Magazine of Business, 131, 132, 145, 175, 247
systematic management, 134–36, 138–39, 163, 164. *See also* Litterer, Joseph; scientific management
systems, 9, 42, 43, 104, 173; cabinet, 75, 77–80, 82; card, 242, 243; compared to individual, 127, 139, 163, 190, 201, 209, 214; definition of, 45, 139; efficiency and, 4, 8, 26, 134, 136, 176, 244; emergence of management and, 135–36; home and, 231, 232, 242–48; office and, 24, 26, 131, 139, 140, 226; record-keeping, 69, 70. *See also* classification systems; filing systems

tabs, x, xii, 3, 7, 18, 22, 72, 105, 115–18, *119,* 120, 256; celluloid, 115, 116, *116,* 117; colored, 117, 165; handling, 16, 115, 116; information and, 14, 19, 25, 27, 164, 173, 182; positioning, 117, 118, 164; visibility of, 56, 107, 117, 119, 127
tabulating machines, 189
Taft Commission. *See* Commission on Economy and Efficiency
Taylor, Frederick Winslow, 17, 85, 104, 158, 231, 242; mechanization and, 176; scientific management and, 136, 138
Taylorism, 158, 224
technology, 3, 6, 25, 33, 68–69, 104,

109, 134, 147, 148, 180, 203, 224, 226, 243, 257; book as, 18, 64, 65, 69, 70; file as, 120, 128, 143; filing cabinet as, xi, 4, 108; folder, 124; gender and, 27, 256; information, 4, 64, 127, 128, 132; materiality of, 15–16; organization and, 140, 148; push-button, 186; storage, 26, 27, 63, 64, 96, 101, 124, 128, 145, 228, 230, 232, 233, 246, 248; work and, 177, 180, 182, 228, 232
te Heesen, Anke, 104, 105, 151
telephones, 8, 59, 149, 156, 178, 189, 218, 256
temporality, 17, 18, 22, 23, 24, 49, 64, 67, 105, 108, 110, 158, 176
tracings, storing, 154
typewriters, 70, 189, 210, 228, 256, 291n2, 291n66

Underwriters Laboratories, 95
usage scenarios, 139–40, 142–45; definition of, 140
U.S. Army, ix

Vawter-Baker, 84
vertical, 4, 24, 27, 39–40, 43, 110, 182, 237, 271n6, 288n29; bias of capitalism, 35, 37; city, 39; defined, 31, 33; efficiency principle, 41, 103–4, 108; structures, 36, 40, 42, 62, 104
vertical file. *See* libraries
vertical filing, x, 31, 70, *107,* 108, 148, 153, 212; base unit of, 120, *122*; critique of, 283n67; definition of, 7, 109
vertical filing cabinets, x, 31, 32, 108–9; and horizontal filing cabinets compared, 99. *See also* filing cabinets
vertical integration, 4, 35, 37, 39
verticality, 23, 110, 120, 128, 237, 264, 271n4; capitalism and, 33, 41, 59, 271n1; churches and, 39–40; critique of, 41–42; definition of, 31, 33; efficiency and, 33, 41, 53,

56, 62; flow and, 36, 41, 42, 59, 62, 104; grid and, 43; mode of organization, 24, 33, 35, 36, 41, 43, 59; skyscraper and, 39–42, 55
visibility: accessibility and, 108, 173, 190, 235; "at a glance," 101, 173; tabs and, x, 7, 25, 107, 114, 115, 118, 127, 157, 164, 173, 182, 230
visible indexes, 147, *149,* 178
Vismann, Cornelia, xi, xii–xiii, 22, 66, 270n43
Von Oertzen, Christine, 295n73

Wabash Cabinet, catalog, *100, 200*
Weaver, Warren, 269n28
Weber, Max, 3, 282n50
whiteness, 197, 198, 199, 206, 220
Williams, Robert, 219
windows, 40, 45, 99, 117; office size and, 44
Wobble Block, 113, *113,* 114
Women's Bureau (Department of Labor), 198
wooden chest/box *(arca),* 170
Woolf, Virginia, 241
Woolworth Building, 12, 37, 39, 40, 42, 273n35; postcard of, *38*
Wooton Desk Company, 49, 233
work: brain, 12, 176, 181, 188, 190, 209, 214; class and, 26, 215, 220, 242, 291n3, 295n73; delegation of, 224, 231, 232; domestic, 186, 232, 242, 243; efficiency, 17, 33, 53, 140, 159, 176, 209, 224, 246, 248; flow of, 36, 42, 47, 48, 104, 235, 273n51; gender and, 4, 12, 14, 26, 58, 177, 189, 195, 197, 201; information and, 11, 46, 218; invisible, 3, 163, 226, 232; knowledge, 159, 179, 180; machine, 25, 26, 143, 163, 175, 177, 180, 183, 185, 189, 190, 201, 204, 220, 294n67; manual, 12, 26, 58, 181, 182, 185, 188, 203, 209, 242; men's, 12, 14, 163, 180, 188, 189, 190; mental, 12, 26, 58, 164, 176, 181, 185, 203, 209; race and, 195,

211, 220, 224, 241, 290n3; scientific management and, 17, 104, 136, 140, 158, 176, 224; sexuality and, 26, 194, 202–3, 220; white-collar, 198, 203, 221, 254, 255; women's, 12, 25, 26, 53, 176, 181, 186, 188, 189, 190, 195, 202, 204, 215, 220, 221, 223, 232, 245, 254. *See also* clerical work; filing; information labor; office work

work environment: character of, 203; digital, 256; middle-class, 204; modern, 10, 43, 44, 59

workflow, *47,* 48, 104, 275n51. *See also* flow

"World's Tallest Filing Cabinet, The," 251, *252,* 253

Wright, Frank Lloyd, 42, 46, 48; culture of congestion and, 41; horizontal/vertical and, 40–41; skyscrapers and, 41, 43; space/freedom and, 41. *See also* Larkin Building

Yates, Francis A., 288n26
Yates, JoAnne, xi, 138, 139, 282n50, 284n20
Yawman and Erbe, 9, 154, 209; catalog, *15, 19, 74, 81, 93, 178, 196*; Direct Name, 166; Efficiency Desk from, *48,* 180, *181*

Zakim, Michael, 188, 291n73
Zandy, Janet, 182

Craig Robertson is associate professor of communication studies at Northeastern University. He is author of *The Passport in America: The History of a Document.*